Praise for *Building Products for the Enterprise*

It's the rare company that achieves product-market fit in enterprise software, and many books have been written about that magical customer discovery process. But until now, there has been no guidebook for the scaling phase that follows. In enterprise software, initial product-market fit only means a growing pile of demands from users, customers, and prospects. In their book, Blair and Ben lay out the key principles of product management that allow companies to conquer the software markets they discover, rather than getting lost in a product maze of their own creation.

—*Andrew Montalenti, Cofounder and CTO of Parse.ly*

Great product managers are master jugglers and expert tea leaf readers. This book is a guide in how to perfect both. I recommend it to anyone looking to learn more about product management for enterprise software.

—*Joe Colopy, CEO/Cofounder of PeopleLove*
CEO/Cofounder of Bronto Software

Building Products for the Enterprise is one of those rare books that really helps you understand a particular job—that of product management. Product management can be the glue that holds the entire organization together when done well, and it is one of the most exciting and entrepreneurial career opportunities. Read this book if you are interested in—and want to master—the field.

—*Brett A. Hurt, founder of Bazaarvoice, Coremetrics, data.world, and Hurt Family Investments*

There are a lot of product management books about startups and how to introduce product management to your company. But where are the books about product management at large enterprise SaaS companies? I was sold on this book when I encountered the accurate metaphor for the job: the Product Manager as a sheepdog. That's who we are—Product Managers trying to steer the herd to the best of our ability on the path toward growth, disruption, and SaaS domination. *Building Products for the Enterprise* is an excellent "how-to" for Product Managers who have found themselves at these growing enterprises where product management is encouraged (and expected) to truly be a business driver.

—*Kathryn Murphy, Vice President of Product Management,*
Salesforce Commerce Cloud

Building Products for the Enterprise

Product Management in Enterprise Software

Blair Reeves and Benjamin Gaines

Beijing · Boston · Farnham · Sebastopol · Tokyo

Building Products for the Enterprise

by Blair Reeves and Benjamin Gaines

Published by O'Reilly Media, Inc., 1005 Gravenstein Highway North, Sebastopol, CA 95472.

O'Reilly books may be purchased for educational, business, or sales promotional use. Online editions are also available for most titles (*http://oreilly.com/safari*). For more information, contact our corporate/institutional sales department: 800-998-9938 or *corporate@oreilly.com*.

Editors: Angela Rufino and Meg Foley	**Indexer:** Ellen Troutman-Zaig
Production Editor: Nan Barber	**Interior Designer:** Monica Kamsvaag
Copyeditor: Octal Publishing, Inc.	**Cover Designer:** Karen Montgomery
Proofreader: Sonia Saruba	**Illustrator:** Rebecca Demarest

March 2018:   First Edition

Revision History for the First Edition

2018-02-26:   First Release

See *http://oreilly.com/catalog/errata.csp?isbn=9781492024781* for release details.

978-1-492-02478-1

[LSI]

Contents

Foreword

Product management is hot! Why? A lot of reasons, not the least of which is the growing primacy of product, where many of our sources of modern competitive advantage are found in products that utterly delight customers. But I also credit Google and the wildly successful product management organization it has built. Perhaps without equal, Google has hired and trained wicked smart people who built some of the most recognizable and delightful products we all know and use. Many are now CEOs or venture capitalists—the power brokers in our current technology ecosystem. Much of the content around product management is driven by this hallowed community. I'm fortunate to work closely with two individuals with this pedigree and have encountered many more, and I'm qualified to say they are the real deal.

Yet, their background is different than mine. Google is a consumer-first company—at least in its early days. Some Google product managers have interacted with sales, but many have not. Much of the idealism that I read about is inspiring to me, yet it doesn't reflect the reality that I've faced trying to be successful in my past. Not all of us live in a world where we can A/B test something in Google search and get millions of datapoints in an hour.

The reality is that enterprise or B2B product management is its own beast. And it's an important one. Enterprise product management is driving our economy. We are designing the future of our human resources departments, our supply chain, advertising technology, data centers, and generally all major infrastructure.

As I read this book, I smiled. I smiled because the stories and ideas resonated so closely with my experience. This is a book written for me. Who am I? I've had the pleasure of being in enterprise product management since I began my career over 20 years ago. I've led product teams, and now I'm the CEO for a company whose customers (and users) are primarily product teams. These are my people. If you're reading this, they're likely yours, too.

But there's no degree in enterprise product management—at least no degree that I've found. I'm very familiar with programs like Pragmatic Marketing and others that have some good training. But, until now, I hadn't found content that perfectly captured the job of enterprise product management.

If you are an enterprise product manager, enjoy this. This may well become our "bible." If you aren't an enterprise product manager, welcome. I'm biased, but I feel there is no better role. For me, it has been incredibly rewarding, an amazing stepping stone for any and all ambitions, professionally and otherwise. Other product people understand what I mean.

Enjoy!

—*Todd Olson, CEO and*
Cofounder, Pendio.io,
Raleigh, North Carolina

Preface

If you're new to software product management, or just looking to learn more about it, you might naturally do a search on the web on the topic. Brew some coffee first, though, because you'll get a lot of stuff back. Some of it will be good; some kind of interesting but incomplete; and a large swath of it, frankly, will be clickbait claptrap. (In other words, a pretty typical breakdown on tech industry topics.) Over the years, Ben and I have taken to sending each other links to particularly egregious examples of bad or amateurish product management advice that we stumble across. One thing we've noticed is that even the "good stuff" is usually geared toward the startup community and consumer tech world, which isn't where we live. (Literally—neither of us lives in Silicon Valley, either.) As a result, as product managers in enterprise software, we've mostly had to figure this stuff out for ourselves.

Enterprise software often seems to live in its own separate universe. This isn't the frothy, startup-VC-industrial complex whirlwind that's covered breathlessly by the tech press. There's a narrative on that side that sometimes suggests that the huge firms that dominate the enterprise market are back on their heels, scrambling to compete with nimbler VC-funded startups that are eating their lunch. Indeed, the joke goes that the VC/startup culture loves to bash "BigCos." Until, that is, a favored startup wants an acquisition. Indeed, the market reality is something quite different than that narrative implies. As of 2014, a group of 10 companies generated 45% of all global revenue in the enterprise software segment. In fact, just five companies—Microsoft, Oracle, IBM, SAP, and Symantec —generated 38% of that revenue (*https://pwc.to/2EX1IMQ*). Just two companies on that same list, Salesforce (#9) and Google (#20), have been founded in the past 25 years. Make no mistake: the overall enterprise software business is very healthy and growing fast, even while undergoing huge changes, like adapting to cloud service delivery and subscription-based pricing models. The enormous growth of enterprise Software as a Service (SaaS) into today's de facto default model for enterprise software represents the culmination of these trends.

Your authors, Ben (Adobe, #11 on the list) and Blair (SAS, #13) have spent our entire tech careers at the forefront of that very shift to the world of enterprise SaaS. The software world we live in is not the hypergrowth, boom-and-bust cycle that we so often see on the consumer side. It's a market of slow ramps, one negotiated sale after another, repeated successful execution, demonstrated business value and organizational collaboration, all sustained over time. Startups are relevant mainly as potential acquisitions, not so often as meaningful competitors. What the startups call their "enterprise" tier is frequently our small-to-mid-sized segment, or even too small an account size for us to touch. This isn't hyperbole, it's just the reality we face in our day-to-day jobs.

Much of the prevailing wisdom about product management in our industry today is derived from lessons learned from the 2000–2008 tech cycle, reapplied to the subsequent one that we're in now. In that era, "hypergrowth" Silicon Valley firms like Google and then Facebook transformed from startups into the giants they are today, and the resulting diaspora of investors and alumni have evangelized the product lessons learned at both companies around the Valley. In particular, Google's Associate Product Manager ("APM") program alone is credited with defining much of today's conventional product wisdom in tech. Product management as a function existed prior to all of this, of course (Microsoft has had its equivalent "program managers" for decades), but it took this cycle of the tech industry's mega-growth for product managers to become a standard feature of software product development.

Yet the more we talked about it, the more we began to agree that there was a real and growing gap between what we read about "PM" in the typical tech circles and what our jobs actually looked like building software products for businesses. Upon examination, we trace this back to the big fundamental business model differences between consumer and enterprise software. The leaders of product at Google, Facebook, and other consumer-facing companies have an awful lot of wisdom, but much of it simply isn't applicable to what we do. The challenges we face in enterprise software are meaningfully unique, and deserve their own dedicated treatment. So that's what we aim to do in this book.

Your authors are both operators. We are both full-time enterprise product managers—not consultants, professional speakers, or "thought leader" VCs. In our day jobs, we figure out how to build amazing cloud software products for some of the biggest and most impactful and influential industries on earth.

Some of our products have almost certainly touched your life in some way. We both love our jobs, and feel like we've learned some stuff that other enterprise product managers, and those who are curious about a gig like ours, might be able to use. So, one unseasonably warm winter night over dinner, we decided that we'd write a book that gave our "insider's" perspective on enterprise product management, and...well, here we are.

This book is for product managers in enterprise software, both those doing the gig now or those who might be interested in getting into enterprise product management in the future. Fair warning: we do not have all the answers, and we don't pretend to. That's also a theme you'll see us return to in this book, so take it to heart. The tech industry is a rapidly changing place, and we know all too well that there's no one, true way to do product management. But we've learned some stuff that has worked for us, and we've learned some stuff that absolutely has not. We're going to push back on some of the hype and airy abstractions of "product," pop a few of the myths you've heard about, and try to give some practical, straightforward guidance that we hope will be helpful to you. We're going to cover, for example, how we decide what products or features to build, how we build a roadmap, how we organize a product team, how we work with other key teams in our organizations, and much more.

We've organized this book into the following, sometimes broad, chapters:

Chapter 1, Why Product Management in the Enterprise Is Different
> What prompted us to write this book was noticing that product management is often discussed in terms that don't really work for those of us in enterprise software. Building software for companies rather than consumers introduces dramatic differences into how our products are planned, designed, built, managed, sold, and distributed, all of which makes our jobs as product managers meaningfully different than for our consumer-facing colleagues.

Chapter 2, Who Are We Building For?
> We start with a fundamental clarification of *user* problems versus *customer* problems. These are different people, and grasping their respective needs is our currency as product managers.

Chapter 3, Three Types of Knowledge for a Product Manager to Seek
> The best way we know to categorize the right domains of practical competence for an enterprise product manager are as organizational, product, and industry knowledge.

Chapter 4, Organizational Knowledge

How enterprise product managers should aim to work with other internal teams, like development, design, marketing, sales, and executives.

Chapter 5, Product Knowledge

How enterprise product managers should approach their product planning process: the product life cycle, building the roadmap, gathering customer input, and measuring success.

Chapter 6, Building Better Products with Data

Product managers often have innumerable sources of data about their products. How to sift through it all to determine which parts are actionable, and in what way?

Chapter 7, Industry Knowledge

How enterprise product managers should understand their market or the context for everything they build. Learn about your market from primary data, third parties, your colleagues and, most of all, the customers themselves.

Chapter 8, The Product Managers

How to get into product management, managing product teams, and making sure you continue to grow in the role.

In addition, we've also enjoyed the great privilege of interviewing several of our outstanding peers in enterprise product management, across a range of companies, to get their perspectives on the challenges of building software products in our industry. We've included these "PM Profiles" at the end of each chapter. And we're not finished! Since publication, we've continued to seek out perspectives from leading enterprise product managers, and we'll be publishing new ones periodically at our book website: BuildingForBusiness.com (*http://building forbusiness.com/*).

We also readily admit (and revisit this in our conclusion) that the way companies make decisions about product strategy, roadmap, and go-to-market are much more complex and diverse than what we can comprehensively cover in these pages. As you read, you might find yourself thinking, "All of this sounds great, but it would never happen that way here." That might be true. Figuring out how to adapt and work these principles into your current organization can be a challenge, but we've seen many product managers begin to implement these ideas, here and there, even in companies with heavy internal inertia. If nothing else, we hope we've given you a way to ask, "Why aren't we doing it this way?" and begin an honest conversation with your peers and management about how you can improve your product focus. The one thing we do *not* want is for you to read this book and become discouraged by the chasm between what we describe and where you think your team currently is. Even in your current situation, we believe you can do great things for your customers. So, hey, keep your head up!

If you don't work in enterprise software, we think you'll still be able to take something useful away from this book. Consumer web businesses often operate very differently than we do in enterprise, in ways that have big implications for how product management is done. That, of course, is the genesis of this book. Yet clearly recognizing the differences between these segments can help each better understand itself.

And if you do work in the enterprise market, welcome. We're part of the same tribe. We understand your goals and, in all likelihood, share them. If you see stuff we've missed in here, we'd love to hear from you. Like we said, we're still learning, too.

Onward.

O'Reilly Safari

 *Safari* (formerly Safari Books Online) is a membership-based training and reference platform for enterprise, government, educators, and individuals.

Members have access to thousands of books, training videos, Learning Paths, interactive tutorials, and curated playlists from over 250 publishers, including O'Reilly Media, Harvard Business Review, Prentice Hall Professional, Addison-Wesley Professional, Microsoft Press, Sams, Que, Peachpit Press, Adobe, Focal Press, Cisco Press, John Wiley & Sons, Syngress, Morgan Kaufmann, IBM Redbooks, Packt, Adobe Press, FT Press, Apress, Manning, New Riders, McGraw-Hill, Jones & Bartlett, and Course Technology, among others.

For more information, please visit *http://oreilly.com/safari*.

How to Contact Us

Please address comments and questions concerning this book to the publisher:

O'Reilly Media, Inc.
1005 Gravenstein Highway North
Sebastopol, CA 95472
800-998-9938 (in the United States or Canada)

707-829-0515 (international or local)
707-829-0104 (fax)

We have a web page for this book, where we list errata, examples, and any additional information. You can access this page at *http://bit.ly/building-products-for-the-enterprise*.

To comment or ask technical questions about this book, send email to *book-questions@oreilly.com*.

For more information about our books, courses, conferences, and news, see our website at *http://www.oreilly.com*.

Find us on Facebook: *http://facebook.com/oreilly*

Follow us on Twitter: *http://twitter.com/oreillymedia*

Watch us on YouTube: *http://www.youtube.com/oreillymedia*

Acknowledgments

No product manager does it alone, even when "it" is writing a book. We are grateful for the support and guidance of dozens who made Building Products for the Enterprise possible. First, we'd like to thank the O'Reilly team: Nikki McDonald, Angela Rufino, Nan Barber, Meg Foley, and everyone else who guided two newbies through the publication process (successfully, if you are reading this). We are indebted to our outstanding review team, which offered countless suggestions to make this text more valuable to you (and also made sure we are not lying to you about how to build great software); this team included product veterans Matt LeMay, Ryan Harper, Sachin Agarwal, Kimen Warner, and Jeremy Horn. We were thrilled and appreciative to have Todd Olson write us the foreword that starts this book off right. This book would not have been possible without Matt LeMay providing general inspiration to us in his book, *Product Management in Practice: A Real-World Guide to the Key Connective Role of the 21st Century*, also available from O'Reilly Media.

Blair Reeves: I've had the opportunity to work with and learn from a great many brilliant and hard-working people, without whose influence this book would not have been possible. The first is the most obvious: I want to thank Ben, who, in addition to being the brains of this operation, is a terrific writer and product manager from whom I have learned a great deal. I'd like to thank many of my old managers for the lessons and experiences, both the easy and hard ones, that have led up to this book—in particular, Chris Benedetto, Alan Bunce, Mike Niemann and Barbara Stilwell. I wish to thank my wife, Laura, for her help keeping me on track, and my dog, Abby, for her reliable looks of faithful encouragement. Lastly,

I want to thank our daughter, Penny, whose due date provided the hard deadline I needed to finish this book.

Ben Gaines: This whole thing was Blair's idea, and without him there is no way this book would have happened. He constantly amazes me with the depth and breadth of his knowledge. Thanks also to members of the Adobe team who provided input on specific topics in the book: Jordan Walker, Brandon George, and Michael Malmgren. The rest of the Adobe product teams—far too many individuals to name—contributed extensively to the long-accrued knowledge and experience that went into much of this book, and I am nothing if not grateful for my teammates (current and former) for their wisdom and support. I'll mention specifically those who have managed me directly during my time in product management, each of whom has been an outstanding teacher and mentor: J.D. Nyland, Bret Gundersen, and Chris Wareham. Last but certainly not least, thank you to my wife, Toni, for her constant encouragement, patience, and love; and to Anna, Eliza, Calvin, and Becca, who are the greatest creations I will ever be a part of.

Why Product Management in the Enterprise Is Different

Even if you're a seasoned product manager, from the business model to the meaning of the word "customer," building products for companies is substantially different than building products for consumers. Here's why.

Product is hot. If you pay attention to the tech industry at all, you know what we're talking about. "Product manager" is one of the most highly-coveted job titles (*http://on.wsj.com/2EMLKrT*) for new MBA graduates. Entire blogs, newsletters, conferences, even boot camps have sprung up around the tech world catering to people interested in the topic. Try putting "product manager" in your LinkedIn profile, and get ready to see a steady uptick in your recruiter spam. There is, in fact, a whole content and social network called Medium dedicated solely to sharing product management thinkpieces.

Yet we still have difficulty explaining to our parents what we do for a living. Neat summaries or the kind of pithy clichés ("the product manager is like a mini CEO!") you often see are either silly or, at best, very incomplete. We're not engineers. Nor are we marketers. We're certainly not "CEOs of the product." But we're the guys who everyone else looks to when there's a question about our products, or "how it's doing," or "what it needs" or "what [it/we] need to do next." And we better have a good answer for those questions, because we're ultimately the only guys whose responsibility it is to know.

In a mature, high-functioning software business, you have a lot of teams hustling around different aspects of a product (or group of products):

- There is obviously engineering, which not only builds what they're asked to build, but tests and patches it for stability, quality, defects, and more.
- In Software as a Service (SaaS) businesses like ours, there's an operations team that monitors the cloud platform and customer tenants and works closely with engineering (hi there, DevOps!) on releases.
- There's sales, the folks who actually breathe life into your business. Sales can encompass everyone from lead development reps, or LDRs (sometimes called account development managers or similar) to account executives, to customer success managers, to sales engineers (or solutions consultants), all of whom are dedicated to acquiring, growing, and keeping business.
- Marketing, whose array of specialty functions can be dizzying: digital, content, demand generation, social, events, brand, product and solution, and more.
- There are "foundational" services like finance, legal, and even IT, that are easily overlooked but absolutely essential to making everything run. And there are plenty more besides.

Right at the nexus of all of these teams—neither above nor below, but working as a peer—is product management. "Product" leads in defining the product vision, establishing the operational plan to get there, and then executing on it. It harnesses the incentives built in to all of the other teams and aligns them toward a single destination. In this way, "PM" is an operational role, not really a technical or specialist one. If you'll permit us just one pithy metaphor, the product manager is a sheepdog; we don't *necessarily* dictate the destination, especially in business terms (an executive team often does that), but we generate a route to get there, a "roadmap," if you will, in terms each team understands.

We're also no one's boss. Product typically works *with* other teams, not *over* them. The product manager is no one's CEO and usually works with little or no formal authority to tell anyone what to do. Indeed, in most software companies, and particularly in enterprise software, one of product management's core functions is this cross-functional collaboration with many different teams to make sure that the team as a whole moves in the right direction.

In most high-functioning organizations, other teams rely on Product to help them do their jobs more effectively. Development relies on product management to define a plan and write user stories, requirements, and acceptance criteria that explain what needs to be built. Marketing relies on Product for product information, value proposition definitions, business case material, market intelligence and, sometimes, content. Sales relies on Product for much of the same, plus determining target market segments, demo cases, answering detailed inquiries, and helping to close deals. Finance and Product rely on each other to build the business through determining pricing, margins, discounting, and so forth. Leadership relies on Product for execution on overall product and business strategy. Across all of these cases, Product is the key index and what keeps all of them churning toward the same destination.

The tech landscape is just so broad, especially when you consider both consumer and enterprise segments, that we doubt if it's very useful to distill the product management function down into just one definition. No two companies will use the product management in exactly the same way. Perhaps being the "sheepdog" doesn't resonate with you. If not, we bet this will:

A friend of ours, Matt LeMay,[1] was recently teaching one of his excellent courses on product management for O'Reilly. In it, he gave his students a quick learning exercise: he separated them into teams, representing different parts of the aforementioned business. He gave each of them certain resource restraints and a finite timeline in which to solve a product quandary. After a period of negotiating with one another, they came back to him, harried, confused, and a little panicked: "There's just not enough time!" they cried. "We can't complete all of these requirements, in sequence, in the time window you're asking for! What do we do first?"

After he explained this ingenious exercise to us, we laughed, because we all got the joke. "Welcome to product management!"

Three Things that Make Enterprise Software Different

There are really three big elements that make product management in enterprise software different: our business model, our specialization, and the split between our customers and users.

---

1 Author of *Product Management in Practice*, available from O'Reilly at *http://oreil.ly/2nQiOzl*.

THE BUSINESS MODEL

In consumer technology, there are lots of revenue models; advertising-supported is obviously very popular, but there are also two-sided marketplaces, affiliate models, direct sales (almost always self-service ones), and much more. In the winner-take-all world we live in today, all of these are fundamentally scale businesses. You have user bases in the millions (or tens, or hundreds thereof) and must make product, pricing, infrastructural, and technology decisions accordingly.

In the enterprise market, we operate almost exclusively on a direct-sales model, whether that means selling good old-fashioned licenses or subscriptions (as we do in SaaS). Enterprise is rarely a scale business. Though there are interesting exceptions, we mainly rely on direct sales to sell complicated products and solutions to a relatively small (compared to consumer software) group of sophisticated customers for large deal sizes. Those deals are negotiated, budgeted, implemented, and monitored for weeks, months, or even years! Sales cycles are long and complicated. (You can probably already see some obvious reasons why it can be very difficult for startups to crack this market.)

PRODUCT SPECIALIZATION

Enterprise software is usually also very specialized and tailored to serving a specific technical or business need. Think CRM, business analytics, or data management. These are solution areas that become very complicated, very quickly. Our challenge as product managers is not only to make these tools deliver more business value over time with new features, capabilities, integrations, and so forth; it's also to understand how our users work, what their jobs and industries are like, and how our products fit into their lives. This is especially true after your product breaks out of the "point solution" category (i.e., meeting a very specific, constrained business need) and grows to encompass multiple different use cases, roles, or industries. (We talk about platforms versus products later on in the book.)

THE CUSTOMER VERSUS THE USER

The last big thing that makes "enterprise PM" tricky is who we're building for. Our users, who might spend a substantial chunk of their professional lives working in our product's user interface (UI), are almost never the same people who actually pay us for it. Our users work directly with our product to perform some function that allows them to do their jobs better, more easily, or efficiently. Our customers, by contrast, are nearly always those users' bosses (or their bosses, or even their bosses' bosses) who write the check.

When you're selling a product that has a five-, six-, or even seven-figure price tag, it's going to be an executive, not infrequently a C-level one, who has the budget juice to pull the trigger. These types of *customers* have a very different set of selection criteria than their direct reports, our *users*, who will actually be using our product day in and day out. Broadly speaking, they tend to care more about hard business results and return on investment (ROI) than a high-design UI and feature checklists. In our experience, they *might* touch the product in question, but they do not live in it day in and day out.

In some cases, your users might exert a strong influence on your executive customers. Executives might short-list a set of potential solution vendors with their department's input, or, often, they might have used a preferred vendor's product before. For scenarios in which the products in question are highly sophisticated and require specialized training or experience, this is more likely the case. Alternatively, there are lots of cases in which executive customers make their choices based on basic functionality and price, with little or no regard for user input (think call center or large retail point-of-sale software).

Very clearly understanding your *user* and your *customer* and how those two interact and influence each other is absolutely vital to being successful as an enterprise project management.

Should you focus on more powerful or flexible user features? An ultra-sleek UI? Expanded mobile support and great notifications? Or do you target additional integrations and adjacent capabilities that can provide "good enough" functionality to allow your customer to end their contract with a competitor? There is probably an infinite list of projects you could spend time on, but choosing which of these two stakeholders to over-serve, and how, will depend on you, the product manager, understanding which is going to best drive your product's success.

Are You Selling to Users First, or Customers?

There are some companies that make a deliberate strategy of conquering their market's total user base community as a backdoor into enterprises of all sizes. A "hearts and minds" route like this is a high-touch and often high-investment strategy requiring a product focus on the user experience (UX), great design, clear features, ample documentation, simple onboarding, and, often, a very heavy marketing spend. By evangelizing the product among subject matter experts (SMEs) and enthusiasts to make it a popular favorite, they're really relying on these communities to indirectly pressure the big companies they work for to "get with the times."

By contrast, other companies are renowned for having a direct line to the executive suite. They rely on brand and solution-level marketing, long-term relationships, executive-focused "thought leadership," and sheer sales savvy to capture business at the C-suite level. (These are the guys whose ads blanket every major airport.)

These strategies are rarely clearly articulated as such, but it's usually pretty obvious which one your company is taking. They are both proven, successful methods for winning a market. Each one has a quite different set of product implications that enterprise project managers should clearly understand.

Clearly differentiating between our users versus customers reveals a layer of subtlety to the problems the product solves, and therefore the value it offers. Whose problems does our product chiefly address? Our users or customers? Striking the right balance of that value will depend on your reading of your market and product strategy. One way to begin that process is to prioritize the problems we want the product to address. So, get ready for a deep-dive in the next chapter into a discussion of whose problems you're really solving.

Summary

Product management for enterprise software is crucially different than for the consumer market. Not only must we differentiate between our enterprise product's user and customer and the needs for each party that must be met, but our product's route to market and the way customers plan for and buy it are completely different from consumer software. Moreover, enterprise software typically operates with a completely different business model than consumer-facing businesses. All of these differences have significant upstream consequences for how enterprise products need to be planned, built, sold, distributed, and serviced.

PM Profile

Name: Mike Niemann

Title: Director of Product Management, DynamicAction

How did you get started in product management?

When a professor and mentor of mine left the startup [Trajecta, Inc.] he introduced me to while earning my MBA at the University of Texas, I thought it would be a huge blow to our fledgling business, but it wound up launching my now 20-plus-year career in software product management. Given the fact that I was Trajecta's first paid employee and supported all aspects of the business (sales, marketing, customer support, IT, etc.), I had a strong understanding of how our core software could be leveraged in multiple other offerings. Accordingly, I stepped in to manage the software requirements definition and development efforts that had been contracted out to a third party.

As a manager, what have been the most difficult parts of your job to learn?

I truly believe that exceptional product managers, even at the executive level, must also retain some aspects of being individual contributors. Of course, the VP of Product must be the strategic visionary, but if you've been a successful product manager in the past, that love for understanding your market, your end users, and articulating the value your offering provides never leaves. As I've ascended from individual contributor roles to being responsible for a team of professional PMs, knowing when to simply let go and allow the talent on my team to execute our vision was challenging. Moving from the small startup environment to a large enterprise at IBM, I made great strides in this area probably more out of necessity rather than true desire. However, I can definitively say the experience made me a better manager of people.

Are there particular backgrounds you think are especially valuable for enterprise product managers?

More than a background in any one discipline, experience has shown me that, first and foremost, successful product managers must have exceptional interpersonal skills. Product management requires the ability to communicate and lead persons with a very wide array of personalities and ways of working. Product managers must lead many people across different functional areas but without any direct authority over those resources. You don't have to be a [computer science] major to effectively work with developers and get them to buy into the product vision, but you do need to develop a sufficient depth of technical understanding to earn their respect. Furthermore, product managers don't need to be successful closers, but they must possess the confidence and poise required of sales professionals. A successful product manager doesn't need to understand all the technical details of deploying JavaScript functions on 10 different web platforms, but she had better know the fundamentals of how data acquisition affects the fidelity of an analytics product. I've hired entry-level PMs who've come from development, implementation engineering, marketing, and other roles within the company. I look for individuals that possess strong product knowledge, understand our market, are throughput machines, and, most important, are well respected by their peers. By possessing those qualities, I can be confident the individual has the tools to succeed in a product management role irrespective of their expertise in any given discipline.

What would you do in tech if you weren't in product management?

That is a great question and from my perspective somewhat of a trick question as the aspect of product management I love the most is the requirement and opportunity to "wear many hats." I've been a marketer one day, a technical writer the next, supporting sales as a product expert, serving as the product evangelist/visionary, and even helping with QA. I tend to get great satisfaction when I can contribute to the success of others. From that perspective, I'd likely gravitate toward a role in marketing. Without marketing talent to help build a sales pipeline, articulate the company's strategic vision, convey tactical offering value propositions, plan events, facilitate customer success stories, etc., development could build the best product in the world, but it might not succeed.

Who Are We Building For?

Product management is always about solving customer problems. How those problems are selected and solved in enterprise software can be significantly more complex than in consumer software, requiring new ways of thinking about the customer, and better alignment between internal teams. But this doesn't need to slow you down.

One of the many things that draws a lot of smart, innovative people to product management is the allure of *creation*. There is joy in creation, in invention. Earlier in our careers, both of us remember looking in awe at the product managers, who, we assumed, had the sole power to decide what features and products the business would create next. At least from a technology perspective, they were deciding how the future of the company would be shaped, and we both wanted to be a part of that. Indeed, this chapter is all about shipping product; the process of deciding what you are going to build, and why; getting it "funded" by your development, product marketing, and other teams; and making sure that your design and development teams understand how it connects to your vision for the product, because that is how you realize the joy of creation.

It All Begins with Who and Why

We now know that "shipping product" is only a part of an enterprise product manager's mandate, and it comes toward the end of a long process of product development that starts with a clear understanding of *who* our product is for and *why* it adds value. These are not simple questions, and, accordingly, their answers are rarely simple, either.

Few people, ourselves included, come into product management with a realistic sense of how the company should decide what to create. Young product managers find themselves inundated with noise, and parsing the signal therein takes practice. Let's count just a few of the demands on their time:

- Feature requests from end users, which are often contradictory
- Demands from sales ("We'll lose this deal unless you build this!")
- Analyzing market trends and drivers
- Buyer/user persona research
- Requests or advice from the development team, such as clearing technical debt, researching new projects or approaches, "science projects," and so on
- Ongoing win/loss analysis
- Partnerships and integration inquiries
- Keeping up with the never-ending amount of industry press and analyst coverage of your market, which includes competitive intelligence

This is only a partial list. Where should you focus? How do you know which topics in each of these areas should be most important to you? On a normal day, your inbox and calendar are likely to be full of opportunities to hear something that someone believes is critical to the growth of your business. Yet, as Winston Churchill once said, *"You will never reach your destination if you stop and throw stones at every dog that barks."*

Applied to product management, the guiding principle here is deceptively simple: *figure out who you're building for, and why.* If your strategy is sound, don't deviate from it every time a customer or sales rep makes a demand. That's how you develop a short-term product perspective, which obviously doesn't produce the kind of innovation that redefines industries.

If we had to boil down product management into two things, it would be these: figure out *who* you are building for by deeply understanding customer problems, and figure out *how* to build by driving alignment across teams.

Customer Problems Versus User Problems

You'd think that "who are we building for" would be the easiest question in the field of product management. We build for the people who are asking us to build the software they want to buy, right? You focus on a bunch of *personas*—fictional constructs that act as convenient representatives for a desired business user with a given role or responsibility—which describe the various industry roles that interact with your product (or who interact with the people who interact with your product), and then you create products and features for those personas. Personas might all have different needs, but within any customer organization, they typically own overarching growth goals for their company. As an enterprise product manager, this is a wise place to start.

So, let's talk about *problems.*

Fundamentally, whether in consumer or enterprise software, good product management involves recognizing and deeply understanding problems. The people who buy and use your software have them. The better you empathize with those problems and internalize them deeply, the better you'll be able to express them to your management, your design teams, and your development teams, and be able to solve those problems by delivering relevant solutions. In Chapter 7, we talk a lot more about how to learn what those problems are. For now, suffice it to say that when you solve the right problems, your software becomes much more valuable, and your business grows.

The reason that understanding these problems is central to knowing who you're building for (and why!) is that enterprise software often involves two different sets of problems: *user problems* and *customer problems.* As we talked about in Chapter 1, the people who use your software day to day are often not the people whose problems are the most valuable to solve (at least, from your business perspective).

Users of your software will often make feature requests or give you a signal that relates to the software itself. For example, more technically adept users will often ask for relatively minor improvements (like user interface [UI] tweaks) that allow them to be more effective or efficient in their daily work. These problems are good to be aware of, and in many cases they are worth solving. For instance, they might make your software easier to use, which helps to expand your user base. Often, these requests will relate to product design or user experience (UX), and this is one of several reasons why some product managers will say that "*user experience design owns the user.*"

But the flip side of that maxim is equally important: "*product management owns the customer.*" By "customer," we mean here the organizations that, to paraphrase Clayton Christensen,[1] hire your product to solve a problem. In short, the customer is the buyer of your software. This doesn't mean the guy in procure-

---

1 Carmen Noble, "Clay Christensen's Milkshake Marketing," *http://hbs.me/2GAQAox.*

ment who signs the contract; rather, it's the executive(s) with purchasing power. It means the group of people who buy your software because they believe it will help them grow their business.

Although user problems will differ from persona to persona and from role to role within an organization (given that two users in the same company might interact with your software very differently), customer problems are common to the entire organization and reflect barriers to your customers' growth. We know we have defined a customer problem when we can finish the sentence, "As a customer of [your product], we cannot grow as fast as we would like because we can't __." If we come across a product idea that can't be quickly and clearly related to a customer's revenue growth, why would we invest to solve it? (Of course, your customer's problem might not be growth, but rather some other goal. We have used growth here simply because it's our most familiar customer problem, and one that is widely shared.)

Take the field of marketing analytics (in which both of your authors have some depth), for example. Marketing attribution, the practice of assigning credit for revenue to the various touchpoints and experiences that led a consumer to purchase, has been a hot topic for most of a decade now. The inability to accurately and meaningfully associate a conversion to the steps that produced it leaves companies unable to efficiently allocate marketing budget and create beautiful, optimized customer journeys, and so these organizations spend inefficiently, wasting large swaths of their budget. (Cue the old John Wanamaker line, "Half the money I spend on advertising is wasted—the trouble is, I don't know which half.")

The clear customer problem here is, "we cannot grow as fast as we would like because we don't have a way to easily analyze and optimize how our marketing spend is actually driving revenue, leading to many missed opportunities." Here are a few other examples of customer problems from our own industry, the kind you often wouldn't necessarily hear from day-to-day users of a product asking for product-level improvements:

- "We have reached a point of diminishing returns from our customer acquisition budget, but still need to hit a top-line growth target of 20% year-on-year. The most promising and efficient way for us to do this is by personalizing the on-site experience."

- "We are facing intense downward pricing pressure from a smaller competitor who is trying to steal market share, and we're are looking for new ways to engage with our own customers to stem attrition."

- "We know that segmentation is critical in marketing, but beyond our legacy market segments that we've been going after for years, we do not know how to find the right groupings of individuals so that we can reach them in the right way at the right time."

- "We have been unable to consolidate our customer data across the enterprise and, as such, have not been able to empower various marketing teams with insights at the brand level."

We believe that a good way to approach your role as a product manager is to treat customer problems as the "currency" that allows you to build alignment around a specific product vision. As a product manager, there is no better way to gain the trust and respect of the groups you're trying to align internally than to consistently and confidently articulate customer problems and why they are critical to the companies that buy your software. Later on, we talk a lot more about how to interview your customers to understand their problems at an organizational as well as a user level, but for now, the key takeaway is that the better you understand what is preventing your customers from accomplishing their goals, the better you'll be equipped to solve the problems that keep their top executives awake at night.

The more you study your market and the organizations (and industries) you serve, the more you will see patterns emerge across companies, and the most salient problems will crystalize for your product management team. A customer problem, multiplied by a group of customers, becomes a *market problem*, and solving market problems is your biggest market opportunity.

It's important to note that none of this should be taken as dismissive of "user problems," which are critical to address because users are also part of the buying (and retention) process. It is, however, a warning against focusing on user problems at the expense of customer problems, because this is very easy to do. It's a classic "squeaky wheel" situation. We will frequently hear from product users about the problems they face, but frequently, user problems have limited upside for you to solve. They might help close a deal or two, but industry-shaping growth comes from solving the big, meaty customer problems that often take years, not months (or sprints!), of work. Fixing user problems can delight your users, which is very important, but they do not necessarily deliver the solutions to your customers' broader challenges that make your product truly indispensable. The best way to become essential for the companies that buy your products is to help them grow their business (i.e., to produce meaningful ROI) in unique ways. That is what solving *customer* problems is all about, and it's as neat an encapsulation of the difference between enterprise and consumer software as we can offer.

Ultimately, all user problems flow out of, and support, customer problems. Solve customer problems in a way that pleases your users, and solve user problems that make their organizations' overarching business goals easier for your users to obtain.

If this sounds difficult to do, that's because it is. The enterprise software space can be difficult for even the savviest veteran product manager to navigate. Mastering the practice of understanding and internalizing customer problems takes practice and effort, and aligning your organization to solve them can be, well, really difficult. But don't worry; later in this book we go into much more detail on strategies that you can use on your first day to help you absorb market, customer, and user problems. But first, let's talk about *how* to solve a customer problem effectively.

From Customer Problems to a Product Vision

To deliver a solution to our customer, we first must align resources internally around what problem we're solving and our idea of how to solve it. Driving alignment across teams in complex organizations is among the top three or four skills that an enterprise product manager must develop.

Creating this alignment is fundamentally centered around understanding the customer problems that we've just discussed. Whereas "who we build for" is all about the importance of finding problems that prevent your customers from growing, "how we address this problem" involves then helping our colleagues in various roles (development, design, marketing, sales, etc.) conceptualize the customer problem and understand the value of solving it. "Who" without "how" is just knowledge; "how" without "who" won't achieve alignment, and it certainly won't produce as effective or as valuable a product.

Here is how we recommend product managers begin to create consensus around an approach to solving customer problems:

1. Define a product vision

2. Establish clear metrics of success

3. Create a core team that is accountable for product success

Let's break each of these down in turn.

DEFINING THE PRODUCT VISION

The single most important thing that you and your team of product managers can do to drive alignment is to establish a clear product vision. This could be the subject of its own book, but the easiest way to think about this is that the product vision is the connection between what your developers and designers do every day and the larger company vision. Every product manager (and really every employee, but especially every product manager) should be able to not just recite the company vision, but also to live and breathe it. As an example, Adobe's vision for some time has been, "Changing the world through digital experiences." SAS's mission is giving customers "The Power to Know." You could start any presentation given at or about Adobe or SAS with these statements, and go from there.

As the product manager, the best way to establish your product vision (if there isn't already a clear one) is to figure out how your product will help your company achieve its vision.

After the product vision is clear, seek to establish the categories of innovation that your product needs to pursue to achieve that vision. You will end up with a hierarchy looking something like Figure 2-1.

Corporate Vision								
Product Vision								
Strategic Element #1			Strategic Element #2			Strategic Element #3		
Key Feature	Key Feature	Key Feature	Key Feature	Key Feature	Key Feature	Key Feature	Key Feature	Key Feature

Figure 2-1. Breaking down the corporate vision into product features

We propose being very clear about this hierarchy, and as you present your ideas to colleagues, go through something similar to but hopefully more visually appealing than Figure 2-1. Remind management how your product connects to the company vision and how the relevant aspects of the product strategy help realize that vision. You'll be all set up to explain how the project you are seeking to undertake brings that product strategy to bear.

Establishing (or simply reiterating) the product vision and strategy is critical to the *how* of shipping product because it will be difficult for your colleagues to put what you're asking them to do into context without the ability to connect the project to the broader strategy and vision. You may have experienced this yourself if you have ever been asked to produce an outcome without knowing why it is important. You know that it can be difficult to stay motivated or to understand the right way to deliver what's expected. On the other hand, when you can clearly see the connection between the project you're working on and big-picture objectives, it's easier for everything else to fall into place. Developers, product designers, and even salespeople are the same way. Clearly defining the goal and the major milestones along your journey is a critical leadership function that product management should serve to bring a product into reality.

MEASURING SUCCESS

Management guru Peter Drucker famously said, *"You can't improve what you don't measure."* This is as true of software as it is of any other discipline. The annals of technology are littered with companies that embarked on what they thought was an industry-changing product strategy or set of features, only to find months or years later that it did not actually affect their bottom lines. If only they had known earlier, they could have course corrected! Alternatively, if their efforts were wildly successful, they could have demonstrated to leadership how effective their work and planning had been.

Thus, as a product manager you cannot consider your software release a success unless you have given serious thought to measurement, which includes both defining the Key Performance Indicators (KPIs) that you expect to improve with a release and ensuring that the appropriate data collection is in place to monitor those KPIs (and, hopefully, to allow you to explain exactly *why* a KPI is moving or not moving).

You will have the most success creating and defining KPIs for a feature or a release in the presence of clear objectives and desired key results stemming from your product vision and strategy. We could devote several chapters to these *objectives and key results* (OKRs), which were originally developed by Intel in the late 1970s, because they are that important to this process. Instead, we will refer you to a couple of the many other sources that exist to help you dive deeply into this topic.[2] The OKR concept is simple and intuitive, but too often ignored by enterprise software development teams. The rest of this chapter is largely about driving alignment within product management teams as well as among everyone else involved in the development process, and there is no better way to do that than to ensure that objectives and key results are clear. Defining the right KPIs for your product relies on beginning with these OKRs first; otherwise, you're deciding where you want to go based on whatever direction the wind happens to be blowing.

There are many different "levels" or "tiers" of OKRs, but here we will focus on product-level OKRs. An "objective" in this context is a qualitative, strategic goal that you and your leadership have for your product during a given period of time. For example, "drive user adoption" or "prepare to enter the cloud market." In many enterprise software organizations, your objectives will be set by senior product leadership and might be handed down to you. In others, you will have the opportunity to engage with your leadership to define the most pressing and critical goals for you and your colleagues to pursue, based on your product vision and strategy.

Because OKRs are designed to help you focus on the most important problems to solve, and because you almost definitely do not have enough design and development resources to do everything you might like to do, it's smart to keep your list of objectives to between three and five. The process of defining and honing these objectives can be grueling, but it sets the foundation for measurement while also ensuring that your software releases are in line with your product vision and strategy.

Key results flow from objectives and, when done well, can help solve the problem of defining KPIs for a feature or initiative. Unlike objectives, key results should be quantitative, and define success for their given objective. You will select them as you look for the things that best set you up for the long-term changes that you want to see in your business. If you were a product manager thinking about the "prepare to enter the cloud market" objective, key results might involve getting a certain number of reference customers up and running on a beta version of a product, winning placement in an industry analyst's published research, and/or achieving a certain percentage of cross-sell with other

---

2 Google, which has famously implemented OKRs to great success, writes about its own process here: *http://bit.ly/2H3VJGw*. Christina Wodtke's *Radical Focus* is also a seminal work on the topic: *http://amzn.to/2Ev2zqa*.

more established products in your portfolio. You can use each of these key results to create a KPI to help define your product roadmap, focusing on the things you can do that will best affect these metrics.

Way too often, this process is done precisely backward. Most enterprise software product teams begin with a roadmap (hopefully based on a product strategy), and then either ignore measurement and iteration altogether or figure out KPIs for each individual feature after it has already been placed on the roadmap. But by starting from objectives and key results and *only then* deciding what to place on the roadmap, you can ensure that what you are building is focused on achieving your most important goals.

It might sound like OKRs prioritize *your company's problems* (growing revenue, expanding the user base, moving into new markets, etc.) when we have just spent several pages trying to convince you to always put *customer problems* first in your thinking as a product manager. In reality, customer problems should inform your product strategy long before you get to objectives and key results, so they are already baked into this recipe. Furthermore, after you know your OKRs, you can use your focus on customer problems to figure out which solutions will drive key results and KPIs the most. For example, if one of your proposed key results is to "drive 20% revenue growth next year," you would take that goal and find out which *customer problems* are most valuable so that, if solved, they would lead to the completion of that business objective. In other words, the best way to achieve the key results and KPIs that you define will be to solve the most pressing customer problems first; if not, you either have the wrong key results or the wrong customer problems.

You might have some projects or features that do not map cleanly to an objective or key result. In all kinds of enterprise software organizations there are executive mandates, customer escalations, and technical debt or infrastructure projects that simply have to be handled, whether they achieve one of your key results or not. There might be smaller features that your development team can do on the side, outside of one of your product objectives. For these you will need to define your own KPIs, and define a plan for measuring those KPIs.

We will discuss that process of measurement a bit more in Chapter 6. For now, consider the following two key points:

- A product or feature cannot be successful unless you have defined and measured success.

- The best product roadmaps are built around the objectives and key results you hope to achieve, not the opposite.

THE PRODUCT LEADERSHIP TEAM

Other books or blogs about product management might have suggested to you that driving alignment around an idea is as simple as teaming up with your engineering and design counterparts to visit a customer together or taking everyone out to dinner. If you're reading this book, however, there is a good chance that your company is a little more complex than the standard-issue startup in which everyone knows everyone and company-wide all-hands meetings can be a simple weekly or daily occurrence.

As a product manager in a larger, enterprise software organization, perhaps with several product lines and competing priorities among different teams, you will need to come up with a way to organize the madness (or, as it's sometimes known in our parts, the "goat rodeo"). Even within a single product team, opinions about what projects should be funded by development will abound. Product managers can and often do interpret the same market signals quite differently from one another—frequently, according to how they affect each PM's own product.

After you have a clear product vision and strategy, as well as a proposed solution to a particular customer problem, how do you make sure everyone understands this connection—and the related business opportunity—the same way? How can you create real consensus so that the various stakeholders can move toward the market in sync with one another? How does leadership speak with a clear, consistent, and accountable voice?

There are many ways to do this, and they depend on a variety of factors. Regardless of whether your organization is a 20-person startup or a multibillion-dollar giant, this process is never easy. Each company is different, but we'll share one method that has worked in several large, complex software development organizations that we've been a part of: the *Product Leadership Team* (PLT).

If your development organization is mature, you likely have a program or project management function, which serves to keep development on track and helps organize distributed work plans. Program management might assist in establishing a weekly check-in among key decision-makers, like development managers, product managers, operations, and product leadership involved in the development and go-to-market process, for example. Each week (or every other week, or every month; you will find the right cadence), this PLT will hear and rule on project proposals, and, because it represents product management, various engineering teams, and others involved in the product development process, it will be able to directly discuss possible projects and reach consensus. Think of it as something of a principals' meeting.

This group will need to be committed to participating, but after the ball gets rolling, it will find tremendous benefit in the form of less wasted effort due to shifting or unclear priorities and more solutions to customer problems making it to market. Ideally, in a larger, more complex software organization, no major development initiatives can kick off—let alone go to market—without first having been approved for development by the PLT. At the very least, this keeps leadership aware of what is going on. When this process is working its magic, it ensures that resources are allocated to the best and most valuable customer problems rather than to what one product manager or another happened to work out behind the scenes with a developer or two and a rogue design resource.

Although the primary goal of the PLT is to choose which projects to execute on, the fact that the PLT keeps leadership informed should not be ignored as one of its key benefits. In many cases, even projects related to the product or the customer experience that do not involve development should go before the PLT so that product leadership can reflect and defend those efforts to senior management.

Depending on the number and complexity of products in your offering, the PLT meeting might be a single weekly check-in for the entire product portfolio, or you might want a separate PLT for each major product division in your group/company. You, our dear reader, might not be in this meeting every week, depending upon your role, but at the very least you will want to include the following people:

- Product management leadership
- Development leadership
- Program management leadership
- Product/technical operations leadership
- Others as needed based on your unique organization
- Others as needed for agenda on a per-meeting basis

As a product manager, you and your development counterparts will get on the PLT agenda as you put together an analysis of, and proposed solution to, a given customer problem. In Chapter 5, we discuss what your presentation to the PLT might look like (although there is no template that will work for every software product, for every project/feature, in every company). You will approach the PLT in the middle of your research to get that group's approval to move forward with design and development. Your friendly neighborhood program manager will be able to help you get time with the PLT as soon as you have established some of the bona fides of your proposed project.

You might be thinking, "At my company, we fancy ourselves agile, but this sounds very waterfall." So, we'll come clean: the PLT *is* largely a waterfall development technique. We'll leave the "Does agile work in enterprise software?" debate for another time, except to say that we think it does, but with some necessary adjustments. Even the most agile development organization in enterprise software has major, often interlocking and interdependent initiatives that require planning and take more than a couple of sprints to complete. This is where the PLT shines. If you are like the companies we have observed, you will also have allowances for smaller, sprint-sized projects that do not need to come in front of the PLT (more on this in a moment); but for projects that take more than a few person-months to complete, getting them sanctioned by key stakeholders gives you authority with development and design teams that can't be faked. Product marketing may also get involved as the projects in question touch upon your communicable value proposition.

This section isn't about how to run meetings, but it is worth noting that even though the PLT can be a tremendous device for driving alignment, that does not mean that *everyone* should be there. Of course, to borrow a line from the musical *Hamilton*, everyone wants to be in "the room where it happens," but the PLT will need to process a lot of information in the form of project proposals, discuss them briefly, and make funding and prioritization decisions. Turning the meeting into a session of Congress will bog you down in never-ending debate and prevent rapid decision-making and alignment, which is the entire purpose of the PLT. This is a decision makers' meeting, not a debate club. Timebox the PLT meeting, mandate that decisions be made in a timely fashion as a result, and abide by Andy Grove's rule of "disagree and commit" (*http://bit.ly/2sor7fE*). At the same time, leadership in the meeting should have enough of their righthand people there to help provide them with context that they might be lacking when it comes to evaluating the value and scope of a project. This is something you'll need to test and optimize for your particular circumstances. Often, some level of delegation of authority is required. That is almost always a healthy thing.

We've seen a couple of examples of how this strategy plays out. In the first, a product team in a large enterprise vendor was asked to sketch out a couple of different ideas for how its platform could integrate with a recently acquired adjacent solution. So, eager to tackle the challenge, this team put heads together with the acquired team's product managers and developed a few ideas for novel new cross-features and integrations that their common users would find valuable. And then they waited. This very traditional, hierarchical organization had no PLT, and product teams were not empowered to make big strategic decisions like this one on their own. Instead, they were required to wait for a green light from multiple levels of leadership that never came. Not only was there no delegation of authority, there was also no time limit placed on making strategic decisions about the future of each product. Eventually, other issues arose and the sense of urgency faded. Only very rudimentary connections between the two products were ever delivered, and this team lost a major opportunity for market leadership.

In the second example, another enterprise vendor had a version of the PLT across a particular solution group, but it couldn't settle on common top-level priorities. For years, the group effectively had twenty priorities—which is to say, it had none. The "PLT" did not have a clear set of overarching goals or a definite strategy to achieve them. This state of affairs finally came to a head when growth stalled and the slow pace of progress across the board meant that nothing was getting to "done." The PLT, feeling pressure from leadership, finally imposed a mandate: product and engineering must deliver X goal by Y date, *or else*. With a sword hanging over everyone's head, hard priorities were finally set and X project was met within the mandated timeline. Even though this "worked," this is still obviously a far-from-ideal priority-forcing process.

Because of the kinds of decisions that a PLT is tasked with making, you might find yourself needing to occasionally take a step back with this group and make sure that everyone still agrees on the most fundamental rubrics for decision-making: your product vision and OKRs. Ultimately, everything that the PLT agrees to do should support that vision or challenge it with new learnings and evidence. Do not be afraid to take time in that meeting to reestablish the product vision and strategy with participants as a way to ensure that, when it comes time to decide on a particular project, everyone is agreed on what you are trying to do in the broader sense and what you are trying to be as a product.

So, we have now defined the PLT: a small group of leaders who make decisions about what, when, and how to "fund" a project with design and development resources, and who can ensure that the software you are releasing supports your product strategy and vision. But we have not yet discussed how to use the PLT to get from innovative ideas to formal presentations, to funded projects, to shipped software. If you're anxious to get to that part, skip to Chapter 5. Before that, though, we need to cover some different ground that goes over some "unknown unknowns" lurking out there for many of us.

Summary

The central challenge for enterprise product managers is identifying *user problems* versus *customer problems*, and clearly understanding how your product addresses each set. Customer problems are your "currency" as a product manager, because it's only by solving them that you prove the potential for your product to unlock transformational value. Addressing those problems requires aligning your organization around a shared vision for the product, which should derive from your company's mission and identity. A great way of translating that vision into an operational plan is via the PLT framework, which you can use as a method to force timely decisions, accountability, and shared goals.

PM Profile

Name: Nick Fassler

Title: Product Manager, Asana

How did you get started in product management?

Way before I ever worked in technology, I fell in love with great software. I was an early adopter of every "web 2.0" tool I could find. While I spent the early part of my career in the nonprofit and sustainability sectors, I always convinced the teams I worked with to adopt early web productivity tools like Harvest and Basecamp.

During grad school, I started pursuing the idea of starting a company. I spent over a year trying to find a sustainability-focused idea to build but ultimately gravitated back to team productivity software. I started Thrively, a tool that focused on improving the performance review and peer feedback process for companies. I raised angel funding, built a team, and launched a product—but ultimately ran out of runway to keep going. That experience made me realize how awesome it would be to build products for a living.

Are there particular backgrounds you think are especially valuable for enterprise product managers?

I know so many excellent PMs with radically different backgrounds, so it's hard to say that one particular background is especially valuable. The most "standard" path to becoming an enterprise PM seems to be coming from a technical background and then getting your MBA. I do think that having solid business fundamentals is especially important for working in enterprise, where you are trying to build for end users and for paying customers. I've also worked with a handful of amazing enterprise PMs that were previously in sales engineering—I think it's a unique role that gives you exposure to both how products are built and the needs of a business.

The necessary skill set to be a great enterprise PM also differs depending on your business model. If you are primarily selling a product with large enterprise deals, the PM role likely works much closer to business functions to prioritize individual customer needs and help close deals. On the other side of the spectrum, at Asana (and previously Yammer), our freemium model means that PMs are often closer to design and user research—we need to focus on building products that end users adopt.

When have you felt most fulfilled as a product manager?

I get so much energy from mentoring new and aspiring PMs. Teaching product management forces me to reflect on what I've learned as a PM and keeps me honest about how I can get better. At Asana, I have been extremely fortunate to mentor some amazingly talented and passionate interns. It's an incredible experience to coach folks with raw talent and drive, and help them turn that into being able to build awesome products. Last summer, our intern built a feature (Asana Templates) that turned into one of the company's largest revenue wins.

What would you do in tech if you weren't in product management?

I'd likely be a growth marketer if I wasn't a PM. I have lots of previous marketing experience, especially in user acquisition, and it's a role that has a lot of impact on how people adopt technology. I'm working to build out the mobile growth team at Asana, so it's fun for me to get to work directly with our marketing teams. I'm often tempted to dive into our Google Adwords ad campaigns, but luckily being a PM keeps me plenty busy.

Three Types of Knowledge for a Product Manager to Seek

Anywhere you go, successful product managers are relentless in their pursuit of knowl-edge, although the kinds of knowledge that matter—and how to obtain that knowledge —vary. We recommend thinking about your research and expertise as an enterprise product manager as falling into three distinct categories: organizational knowledge, product knowledge, and industry knowledge.

Congratulations on your hire or promotion into the role of product manager! If you're like most new product managers, you're full of energy and zeal, ready to conquer the world through sheer grit, determination, and lots of sugary treats or booze carts for your developers. (Blair prefers bourbon and gummy worms, whereas Ben goes for ice cream. To each PM, their own.) You're instantly going to win over your colleagues and ship great product features with all of the pas-sion you bring to your company. Right?

Well, sort of. Zeal *is* an important aspect of successful product management. When you're communicating a vision and arguing to resource your roadmap in a particular way that supports that vision, passion is critical. This isn't to say that you should be standing on (or throwing) chairs, or raising your voice, but believ-ing in your own message and making it clear how much you care about it is key. When everyone in your organization can see how strongly you believe in what-ever you're trying to get built, it's much easier for them to rally around you than if you make an emotionless but correct argument. One of our favorite quotes about a product manager from one of his colleagues was, "When I see how passionate [our product manager] is about our product and company, it makes me excited for our future."

But the energy that you bring to your team as a new product manager isn't enough. We have seen new product managers—new to the product management function, a company, or to an industry itself—who have overextended themselves on passion alone or lacked the concrete knowledge required to be truly effective and, more important, failed to develop either. The danger in this common scenario is that a product manager's influence needs to be wielded selectively, and zeal without knowledge can lead to disaster, for your company, for you, and, perhaps most important, for the trust others place in you.

Imagine that you, as a new product manager, are put on a highly visible project and told to make it go. You can rally the engineers and the sales reps with your passion, but how do you know that you're solving the right problem in the right way? It is tempting to assume that your customers must be like you, or that you can guess the things that keep them up at night, or that you will be able to drive the project to success on the back of your own energy. You might be right some of the time. Remember that in your role as "key influencer," the ability to bring both zeal *and* knowledge to bear on your audience is critical. But without putting in the work to obtain actual knowledge, unnecessary failures will litter your path.

So, what knowledge are we talking about? Glad you asked.

Three Categories of Knowledge

You've got zeal in spades (we hope!). So how should we think about acquiring the knowledge necessary to ship fantastic products and ensure that they solve actual customer problems?

We propose that you approach this question by dividing it into three categories:

Organizational knowledge
    This refers to understanding how your company *really* works. Because we need to get the right people behind every major decision we're pushing, we need to learn who those people are and about their motivations and incentives. Don't assume that you know all of this already. This kind of institutional knowledge tends to be vastly more important for enterprise software companies, which tend to be bigger, older, and perhaps more bureaucratic.

Product knowledge

Product knowledge means knowing the ins and outs of your product: its limitations, its benefits, what users love about it, and what they hate about it. Techies tend to fetishize all the minute technical details of their products, and that's not necessarily what we mean here. Rather, it's more important that we know enough to be able to empathize with users to help create solutions to their problems. If we don't understand the product, it's difficult to empathize with end users who do.

Industry knowledge

This is arguably the most important of these three areas of knowledge because it represents a deep and thorough understanding of customer problems that remain unsolved (or inadequately solved) in your market, or adjacent/similar markets, today. It's through this domain that we recognize the largest market opportunities.

Over the next several chapters, we dive deep into what each type of knowledge really means for you as a product manager. For now, let's just briefly explain each area of knowledge and discuss how we can obtain them.

ORGANIZATIONAL KNOWLEDGE

As we've discussed, it is often very difficult for startups to crack the enterprise software space, meaning that the dominant enterprise software vendors tend to be large organizations. As a result, you might be one of hundreds or thousands or even hundreds of thousands of employees who contribute to your company's success, each in their own way. It can seem like a maze of directory listings impossible to parse, even if you've been at the company for years, let alone if you're a new employee. How does an individual navigate this landscape?

The simple answer is that an enterprise product manager does (almost) nothing alone.

Want to ship a feature? Great. At minimum, you'll need to align engineering, operations, sales, support, design, and marketing (both demand generation and product marketing, in many cases); often, you can add accounting and finance as well, not to mention the international teams that frequently operate more or less independently from their counterparts at headquarters.

Do you have a critical customer escalation that you need resolved today? Okay. Who has the answer? The documentation team? Support? A partner? Where do you go to find immediate help?

A simple example from our own experience illustrates this last scenario. At Adobe, Ben frequently is asked time-sensitive questions from sales or account teams. In many cases, he doesn't know the answer to the question. And that's okay! Rather than throw up his hands in frustration and risk unnecessary delay or even losing a deal or customer, his knowledge of the organization is such that he invariably knows the right person, or at least the right team, to answer the question. Much more often than not, Ben is able to steer the request to the best possible respondent within a matter of minutes. It means more than merely knowing that there is a support team or a documentation team. It means knowing which member of those teams has the information we need, which members would be willing to get involved in the situation, and which members are good enough at communicating sometimes tricky answers to potentially introduce them to a customer directly.

In another example, a product manager we know set out to improve the customer experience around implementing a particular software solution. The engineering team with which she was most closely aligned didn't have any available resources to work on the problem. But by interviewing colleagues who had a bit more organizational knowledge, she learned of two engineering teams *halfway around the world* whose incentives and motivations happened to be perfectly aligned with what she was trying to do. These teams, which would have been completely invisible to someone without any organizational knowledge, turned out to be a much better fit to work on the project than her own group of developers.

How does one obtain organizational knowledge? In many ways, this is the most difficult of the three types of knowledge that we will discuss, because it can simply take time. Not only does it require an understanding of the (often international and frequently poorly documented) organizational structure, but it also requires a sort of social graph: who works on what projects, who knows what information, etc. More subtly, it requires you to understand motivations and incentives: what does each individual (and team) want, and why? That can be almost impossible to learn without experience, plain and simple. Organizational knowledge is often a function of tenure. But there are some simple strategies for speeding up the process that we will suggest.

PRODUCT KNOWLEDGE

It should go without saying that it's difficult to be truly successful as a product manager without a deep knowledge of the product itself.

There are primarily two nonobvious reasons to pursue product knowledge. The first of these is *empathy*. Let's look again at the example we just presented in which a product manager was seeking to improve the implementation experience for a particular product. How much more powerful would her arguments and design influence be after she had actually implemented the product herself, and gone through the steps that she was trying to improve? How much easier would it be for engineers, sales reps, marketers, and others to trust a product manager who understood from personal experience what works and what doesn't work for users?

Empathy can be the most important skill for a product manager to develop, because it is what makes the difference between *thinking* you solved a customer/user problem and *actually* solving that problem. Without it, you will find yourself skewing toward solving your own problem, not your customer's problem. Among the very best ways to develop empathy is to put yourself in your customer's shoes and evaluate your product, learning its benefits and shortcomings.

The second reason to pursue product knowledge is *trust*—specifically, earning the trust of customers and the trust of your development, sales, and marketing teams (among many others).

Product knowledge does not mean being a walking encyclopedia of every minute detail of your product. We have always found customers and internal teams to be understanding of the fact that, especially in enterprise software, where extensive customization is so common that no two deployments behave exactly alike, product managers might not have all of the answers. We frequently have customers teach us a thing or two about our own products!

That having been said, both customers and internal teams need your help in order to use and sell your product, respectively. When customers see that you understand your product deeply, and can relate both general information and specific use cases that speak their language, they become much more willing and likely to open up and share with you their problems and concerns, which is perhaps the most important input you can receive as a product manager.

Similarly, internal teams provide you with market insight that can be difficult or impossible to achieve any other way: trends they are seeing in the deals they are working on, or new competitors sprouting up in your space with compelling offerings. Your ability to contextualize these data points goes a long way toward building trust among those teams, especially as it relates to combating misinformation with accurate product information. Without product knowledge, you might not be able to answer the questions "Can our product do that, too?" or

"How should I interrogate what our competitor is claiming?" When you answer these questions accurately, based on your product knowledge, you begin to win the trust of groups that can become your best allies in creating winning products in the future.

Fortunately, product knowledge is the most practical of the types of knowledge discussed. You can obtain it in a number of ways, and (depending on the complexity of your product), often in a matter of weeks or months rather than years.

INDUSTRY KNOWLEDGE

Although organizational knowledge and product knowledge are tremendously valuable, we have to admit that if we had to pick one type of knowledge to be successful in product management, we'd pick industry knowledge. Of the three types, it is most directly associated with the ability to deliver successful products that will grow your company's revenue.

Industry knowledge breaks down into a few categories, but generally we mean having a subtle and multifaceted understanding of the trends that drive buying decisions in the marketplace where you compete, the up-and-coming solutions to current (and future!) customer problems, and how those customer problems vary by industry in your customer base.

In short: know the problems your customers are facing, and know why they are valuable problems to solve. *Customer problems are the "currency" of product management.* The better you understand them, the richer you are.

To illustrate the importance of market knowledge, you need look only as far as a product backlog. If you have one already, we'll bet that it has far more on it than you can possibly solve with the time, engineering resources, marketing resources, and so on available to you. If you do not have one yet, imagine a list of 200 product features or process improvements, all of which are valuable. Now imagine being asked on which five of those features the company should focus.

If you understand what is driving purchase decisions and what problems keep your customers' C-suites up at night, prioritizing that backlog won't be as much like picking which baby you love the most; you'll need to make some cuts, and there might be some choices that are roughly equivalent, as well as some downrange trade-offs to consider, but it can be done. Without market knowledge, you may as well throw darts at a dartboard. (Just remember to keep the bourbon for *after* that meeting, please.)

Industry knowledge is what allows you, as a product manager, to write a Market Requirements Document (MRD), which sets the vision for your product, lays out the core tenets of the product strategy, and essentially answers the question "What is the market in which we operate going to require of us in order for us to grow our business?"

Perhaps because it is the most strategic of the three types of knowledge discussed in this chapter, it can also be the most difficult and time-consuming to obtain for those who are new to the given market. (If you are moving into a product management role from within the same industry, you might find this easier than someone who has been a product manager previously but in a very different space.) In Chapter 7, which covers industry knowledge, we talk about how you should get started on it, as well as several things that we wish we'd have known at the beginning.

You might have heard it said that product management is mostly a "soft skills" position. Among the many soft skills required, passion and knowledge are two of the most fundamental. Without both, a product manager will be hard pressed to influence the kind of change that he is uniquely positioned to enact. To some extent, passion is something that is impossible to teach. If you're not excited about the business you are guiding, you might want to find a new business. You can, however, obtain knowledge.

It Starts with Communication

The product manager is ultimately responsible for bringing disparate teams together, getting them bought into her product vision, and keeping teams (and often customers) aligned both before and after the release of a product or feature. Ensuring that this process goes as smoothly as possible, often in large and complex organizations, requires great (not just good) communication. Although there are many skills you will want to develop as a product manager, those related to communicating ideas will be among your most valuable; in fact, communication might be *the* most important factor leading to success in enterprise product management. And because great communication can be a bit more challenging in this type of software company, in this section, we give you some guidance to help you to get started.

The term "communication" can mean all kinds of different things. We are not talking about communication narrowly, such as the ability to write a coherent email or to execute a decent roadmap presentation to a customer, although those are both part of it. More broadly, we mean the ability to communicate difficult concepts clearly in different ways to different functional audiences (development teams, sales teams, marketing teams, customers, industry analysts, etc.) via different mediums.

Product management sits at the crossroads between a dozen different internal teams as well as customers, partners, and other external entities. Nobody is better positioned than you are to bring these teams together to achieve something great. But each of these groups (including product management itself!) has its own incentives, goals, hopes, fears, habits, traditions, and processes. This means that not every group responds to the same stimulus in the same way.

Suppose that you want to get your sales and development teams to buy into the vision that you're setting for your product. The way most enterprise salespeople are compensated, they are typically focused on short-term bookings objectives, living quarter to quarter or year to year. (And, by the way, this is actually a good thing.) Your product vision matters to them primarily because it's what gets prospects excited about their future relationship with your company; but how difficult it will be to achieve that mission matters much less, provided you have *something* on the roadmap that they can talk about. Great engineers (the ones you really need to buy in to your vision) are, by necessity, more pragmatic thinkers who will immediately wonder how to get from point A to point Z in a reasonable timeframe. Your vision matters to them because it frames how they will solve problems along the way toward achieving that vision, and when they believe the vision is realistic, they will apply their understanding of it to every problem that they solve.

If you are communicating this vision to your sales team, you will often do it by talking about it in the broadest possible terms and pithy statements that sound amazing but leave the door open for them to explain it however they want. You will likely talk about how unique this vision is and how your company is the only one capable of achieving it. You might even call out specific competitors and explain why your vision will put them in their graves.

If you're communicating that same vision to the development team, you will likely boil it down to specifics: *how* we are going to achieve the vision, and *when* we are hoping to reach checkpoints along the path. Your development team probably doesn't care that much about the competitive landscape (nor should they; it can benefit them very little), or why your company is uniquely positioned to make this vision reality. They care that it is the right thing to do, that it solves an interesting problem, and that it makes sense.

The way you have these conversations (and others like these) will often be completely different, and what resonates with your sales reps can be Sanskrit to your engineers (and vice versa). The questions you get from each group will also reflect that group's experience, knowledge, goals, biases, preferences, and so on.

Note that it's the same core message in both cases, but the way you convey the message might be wildly different from one audience to the next. But, of course, you cannot succeed without these (and other) groups being aligned and bought in. And it's primarily up to product management (with the help of other groups, like product marketing) to create that alignment. Thus, you need to become great at tailoring the message to the audience.

Throughout this book, we return to best practices for communicating with different teams throughout your organization: development, marketing, sales, leadership, and more, because communication is such a vital skill to any successful product manager. But let's begin with some quick tips for the group that matters most: customers.

Communication isn't just for internal teams! As a product manager, you should be spending a considerable amount of time speaking directly with customers (and when we say "customer," we include prospects and former customers in this group, as well). Communicating with them is different for obvious reasons: they aren't part of your company and have their own very different incentives, motivations, and perspectives. In fact, the scope of your customer communications is so broad that we cannot cover all of the possible situations in which you might be "in front of" customers needing to figure out the right way to speak and listen. But here are some general guidelines to frame how you approach these interactions:

Become comfortable with different industries

One of the best things you can do when speaking with a customer is convey at least a basic understanding of the challenges facing them in their industry. Your software might be targeted at a specific industry such as financial services, which makes this a bit easier (although every industry has subindustries within it), but if you make general-purpose software, familiarizing yourself with each industry's market drivers, challenges, and key use cases will help customers trust that you have their needs in mind as you are designing and shipping software. It also helps them grasp the value of what you might be sharing with them. Making a publisher think about how to apply a story you tell about an ecommerce brand only creates a barrier to value recognition. You don't need to be a true thought leader in any one industry, but speaking the language will go a long way toward building trust.

You will overestimate how much your customer understands your company

Remember that most software companies look healthy from the outside. They tell stories of amazing things that their customers are doing, and publish press releases when software ships. From where you sit, you might see nothing but dysfunction and decay, but your customers cannot see "how the sausage gets made" unless you tell them. Even when you are feeling discouraged and desperate, convey optimism and confidence in your communications with customers.

Never blame another internal team

Along those same lines, there might be times when something goes wrong (a product bug, a missed ship date, an unexpected limitation, etc.) and you are tempted to lay it out for the customer the way you see it. Transparency is good—in fact, it can be one of the best things you can do in your interactions with customers—but blaming another internal team for whatever is frustrating the customer is not. It suggests to the customer that your company is not united and snipes at one another rather than coming together to do what is best for your clients. Even when another person or team really is at fault, presenting a united front to the customer means either refusing to assign blame or creatively "hoarding blame" for something that might not have been your fault.

Be honest and forthright about your product's limitations

This one might sound counterintuitive; we want the customer to buy the product, don't we? Then why would we talk about limitations? Enterprise customers want a partner in whom they can trust. Salespeople have an obvious agenda, which is to sell software. You have an opportunity to represent yourself as someone who is open, who isn't coin-operated, and who is looking out for the customer's long-term best interest. The thing to remember here is that your customer is likely to find out your product's limitations anyway; an immature (but common) approach to sales would be for the rep to say that he will have booked the sale long before that happens. If nobody else is going to do it, you need to be responsible for the long-term relationship. Being clear and straightforward about what you produce adds to your credibility and sets the stage for your customers to be successful with your product.

Summary

All good product managers bring enthusiasm into their roles, but the most successful ones are those who manage to quickly develop three domains of knowledge: organizational, product, and industry. These three domains encompass a wide range of understanding about the companies we work in, the products we build and sell, and the industries into which we sell. Some of these are easier to develop and more critical to success than others, but it is at the confluence of all three that true product success lies. The next several chapters explore each area individually.

PM Profile

Name: Aubrey Rupp

Title: Senior Offering Manager, IBM

How did you get started in product management?

My background was in business and technical analytics consulting. After years of implementing products and helping customers overcome product and business challenges, I wanted to become part of the solution side. By getting into product management, I decided I could apply my field experience and customer use cases to craft solutions for customers, to improve our products and drive business impact within our organization. As a kid, I always loved math problems, especially the hard ones, because a) they were solvable with the right level of focus and understanding, and b) the euphoric feeling that comes with getting it right. Product management isn't much different. Fewer quadratic equations, but still finding the right approach to solving the problem and the feeling of success if you get it right.

What have been the most difficult parts of your job to learn?

Getting the right users involved and collecting the appropriate user feedback on an idea. It's very easy to think that most users work and think like I do. When coming up with a solution to a problem, it's easy to just implement based on what I think I know. The hard part is making sure that features and solutions get users' eyes on them early—long before the code is written—so that we can be sure we are solving the right problem and solving it correctly. Balancing the needs of the users and the market. It's important to know both, and know them well. In an ideal world, these two want the same thing, but it doesn't always work out that way. Trying to advance the market while staying in sync with the needs of the customers is a constant battle. Too much emphasis in one area can stall market growth, and too much in the other can create frustrated customers. It's about finding balance.

Are there particular backgrounds you think are especially valuable for enterprise PMs?

As an enterprise PM, it's crucial to have deep knowledge of the product, of the customer, and of the industry. Coming from a background of implementation and business consulting, I believe that deep product knowledge can be a cornerstone of success in product management.

Once in the product manager role, a person can learn how to write a user story, Agile, or how to run a scrum. The in-depth knowledge that you get from using a product or helping a customer solve a problem with your product can really jumpstart your career. Coming into product management with customer relationships and strong knowledge of use cases can accelerate your ability to take your product to the next level. One word of caution—too much customer focus can cause you to take your eye off of the big picture strategic initiatives, so understanding this balance and need to balance strategic and tactical is crucial. Regardless of background, passion for the industry is an important characteristic of any product manager.

What would you do in tech if you weren't in PM?

If I weren't an enterprise product manager, I'd consider going into user experience design. Not because I'm a strong designer, but because nailing the user experience is a key part of making any product successful.

If I were to take a completely different approach, I'd move away from corporate software development and into customer strategy at an agency. This would give me the opportunity to engage with many different customers of all shapes and sizes and work with them directly to achieve business results. I'd like the opportunity to craft solutions to individual problems instead of trying to solve at such a massive level.

4

Organizational Knowledge

Enterprise product managers work with a dizzying array of teams and roles, from engineering and design to sales and marketing (and more!). Knowing how to get the best work out of each group in an enterprise software context takes practice, but is a key skill on the path to product management success.

If you play any sort of musical instrument, you already know that playing solo is an entirely different ball game than playing with other people. Although some amazing music was only ever meant to be played unaccompanied, pieces for a string trio or quartet can be positively magical. In a small chamber ensemble of three or four, personal player differences matter tremendously. Each member needs to know *exactly* what the other musicians are doing and the minute variations in how they tend to play in order to adjust their own tone, volume, or tempo at the slightest cue.

Yet precious as it might be, no quartet can ever do what a full orchestra can to convey the majesty of a great symphony. In an orchestral setting, individuals have a bit more margin for error (unless you're first chair—in which case, go practice). Yet the trade-off is that a good orchestral player needs to keep track of many more parts, too. These players need to know what the first violins, violas, percussion, or winds are doing in the next few measures so that they can prepare their part or entrance. They must adjust their volume to account for the cellos (who are, of course, the backbone of any orchestra[1]). If you come in a quarter note before the timpani at a *forte*, well...that's going to be an embarrassing night.

---

1 Full disclosure: Blair is a lifelong cellist.

There are many parallels here to product management in a mature software company. At its best, a software organization that is firing on all cylinders and executing perfectly closely resembles a world-class ensemble like the Berlin Philharmonic or Chicago Symphony Orchestra, whose many different parts harmonize and support one another. Product management (just like the cello section) is a critical function in making sure this entire organization is aligned—if not toward a common goal, then at least toward mutually interlocking and supportive ends.

Day-to-day collaboration with five stakeholders in particular make up a *sina qua non* of enterprise product management: development, design, executives, marketing, and sales. Simply put, developing effective and value-adding working relationships with these five groups of people will make or break your success as a product manager. Over the next several sections, we delve into the organizational knowledge you need to work with each of these teams to make your product successful. In each section, we also share some tips to help you communicate more effectively with these groups.

Development/Engineering

There is arguably no working relationship more important in the organization than between product management and development/engineering. (Note: in the interest of brevity, we refer to this group simply as "development" from here on out.) It is development's job to build the thing; it is product management's job to explain to development what to build. One without the other is either useless or flying blind.

The independence of development teams tends to vary greatly depending on the size and maturity of the organization. A small startup, for example, might have no product management function at all, and just one or a handful of developers. The CEO or CTO or head developer is effectively the product manager in this case. As the company grows, however, the sophistication of the software product tends to increase nonlinearly, requiring developers to be much more focused on engineering and maintaining the solution and less on the sprawling business, market, and customer demands that product management steps in to address.

Thus, in growing companies that have emerged from the startup category and graduated into a growth stage, defining roles and responsibilities between product management and development should help make the latter's job easier, more focused, and better aligned with maintaining a high level of productivity.

This transition requires a lot of trust between both teams—the last thing anyone wants is conflicts over turf and responsibilities. One excellent and frequent solution in this case is to promote to product management from within the company.

More mature companies have typically established a clear standard operating procedure for how product management and development work together. These will often rely on a common paradigm for software development like Agile/Scrum or a Waterfall model. We're not going to delve into the pros and cons of these frameworks here—not only have oceans of ink already been spilled debating them elsewhere, but the applicability of different models is highly dependent on contextual specifics that teams in the field know much better than we do. Suffice to say, we trust that development and product management have already wisely chosen the best software development process for their particular situation. Nevertheless, new product managers who aren't familiar with these frameworks already are strongly advised to become acquainted with them, at least.

No matter which framework you use, you'll need to begin with a basic idea of what you want to build and where you want to get to. In Chapter 5, we discuss how to use market and customer input to build a product vision and strategy; for now, we get into the nitty-gritty part of putting that strategy into action. Rubber, meet road.

STRUCTURING WORK WITH YOUR DEVELOPMENT MANAGER

To scope and plan work clearly, we use the Agile concepts of *epics*, *stories*, and *requirements*, which are fairly standard in the enterprise Software as a Service (SaaS) world. An epic is one large collection of stories which should, in some meaningful way, represent encapsulated, material progress toward your broader product vision, whereas stories often constitute incremental steps toward completion of an epic. At the beginning of sprint planning, you should sit down with your development team(s) to go over the epic(s) in question and make sure each member of the team understands the value it represents to your customers (and, if applicable, your users, too). Only then can you begin the process of decomposing that epic into individual stories, and the requirements that might make up each one.

Before we get to divvying up those requirements, let's talk for a moment about the person who actually does that: your development manager.

The development manager can either be a designated senior developer or, in more mature teams, a person with a specific competence and sole focus on engineering project management. *We cannot stress strongly enough how great an asset a strong development manager can be for your development team and your company as a whole.* The best development managers we've worked with provide a few highly specialized skills:

- They intimately understand the technical strengths and weaknesses of each individual developer on their team, which allows them to more effectively assign and structure work to be done.

- Experience allows them to accurately estimate the level of effort (i.e., "story points") required to complete work on the backlog.

- They can raise caution or red flags early about problems anticipated or encountered, and help brainstorm workarounds.

As a conduit of team-specific knowledge, your development manager makes effective planning possible over the course of many sprints. With accurate story point estimates and work logging, you can derive better velocity estimates—basically, a measure of how much work can be completed within a given timeframe—which enable process improvements as time goes on.

When it comes to assigning requirements, defer to your development manager's judgment. Product managers are typically more helpful in answering developers' one-off questions that arise about specific requirements (e.g., "Should this list show full descriptions, or excerpts?") than in getting deeply involved in details about the technical specifics. This distance can be difficult for ex-developer product managers to keep, but ultimately allows development to be more efficient at its job.

How Technical Should Product Managers Be?

This touches on a common debate in product management circles: just how technical should product managers be? As we've mentioned before, our view is that product management in enterprise software is an operational, cross-functional role, of which technical know-how is only one part, and not the most vital. Even if the product manager *is* an experienced developer, most enterprise software involves sprawling, intricate code bases, often with spotty or incomplete internal documentation that few, if any, individuals fully understand in its entirety. To go mucking around in that world, a developer needs to tread gingerly and possess the understanding of what he's doing that comes from living in that code base day in and day out. That does not describe most product managers, which is why our recommendation is that most product managers should not be touching code. Of course, there can be certain types of product manager roles for which direct code base contributions are appropriate, but in our experience, those are few and far between.

WHAT TO DO WHEN THINGS DON'T GO ACCORDING TO PLAN

When you hear the word "engineering," some of you might think of Montgomery Scott ("Scotty") or Geordi LaForge, both engineers from the *Star Trek* universe. On the fictional *USS Enterprise*, Engineering's job was often to assess risks, play it conservative, and say no when Captains Kirk and Picard (respectively)—much like product management often does—asked them for the impossible. Seemingly in every episode or film, the captain would thoughtfully push back on engineering's concerns, and engineering would pull a miracle out of its hat. Working cooperatively, engineering and the captain save the day.

Just like Scotty or Geordi on the *Enterprise*, your own engineering team has a natural tendency toward conservatism. Developers would always (and understandably) rather take more time rather than less for prototyping, testing, QA, and so on. This is where product managers sometimes must step into an investigative role during a sprint and ask why certain stories are taking longer than expected, why velocity is down, and sometimes reprioritize on the fly so that critical work is finished before less-critical stuff. Just like before, the development manager is a vital partner here who can explain what new issues arose that slowed down progress. Some delays and unforeseen obstacles are just inevitable, and communicating them early helps to set expectations for that sprint's outputs. Other times, understanding why certain stories or requirements took longer than expected can lead to avoiding similar delays in the future, which is a good use of a post-sprint review.

Yet, inevitably, one day you'll be nearing the end of a sprint when you see that a high-value story or two look to be at risk of slipping out of the release. You'll go talk to your development manager, and she'll explain the reasons why, but the long and short of it is that the story is at risk. What do you do?

The answer will, of course, just depend on your situation. But here are two *competing* considerations that might help you think through this dilemma. On the one hand:

Technical debt can be a tool

A lot like consumer debt, technical debt—i.e., putting off important engineering work that might not have immediately obvious customer value, such as refactoring inefficient code—should be avoided wherever possible, because it will always catch up with you and usually cost more to resolve in the long run. Yet also like consumer debt, technical debt can be an effective tool for solving short-term problems. In enterprise software, we often telegraph certain features or upgrades to customers in advance, and when we don't deliver them, it creates bad will and hurts our credibility in the future. *As a general rule*, if given the choice between saving our credibility with the customer and taking on some technical debt, we will take the technical debt seven days a week, with a double serving on Sunday. Typically, technical debt can be resolved a lot quicker and more cheaply than building back a customer's trust.

And on the other hand:

Rushing out bleeding-edge features is risky

"Move fast and break things" is a recipe for disaster in enterprise software. With the obvious exception of early-adopter or beta user scenarios, rolling out lightly tested, "bleeding-edge" features without adequate QA and feature hardening runs a high risk of delivering a bad experience for your users, who will wonder why they're paying you so much money for an incomplete product that they can't trust with their business. Enterprise buyers don't want half-baked solutions; they expect, and frankly deserve, a well-engineered product.

No, we aren't being cute. Both points are true, and like any interesting problem, they will require application within your own specific context. Some questions you should consider are: How important *are* those high-value stories? Are customers waiting on them? Can you roll them out with reasonable confidence that they'll work as intended? Does sales need them to build pipeline? What work will have to drop out of the sprint, or what technical debt will you incur because you chose to finish those stories instead of other items? Only after talking through these questions with your development manager can product management decide on an informed course of action.

COMMUNICATING WITH DEVELOPMENT

Helping development understand customer problems deeply enough that they can innovate to solve them is worth every minute that it takes. Here are some tips to help you think about how we communicate with our development counterparts:

Help them grasp the problem and then let them figure out how to solve it
This might mean putting them in front of customers directly, so that they can see or hear customers' pains with their own eyes and ears. Where that isn't possible, it means communicating back what you're hearing from customers, prospects, and the market. But beware of being over-prescriptive of describing solutions to problems; you likely work with engineers who are great at solving problems with technology after they understand the problems and use cases at hand. Engineers' solutions, in most cases, will be better and more complete if they are allowed the freedom to innovate without being told exactly how many columns should be in the database or how many pixels high the button should be. If you are overbearing in your communications with engineers, do not be surprised if you get exactly what you ask for instead of much more.

Hold them to timelines and commitments
Engineers often tend to want to solve the problem comprehensively, and the more they learn about the problem, the more they see things that should be added to the solution. This is the dreaded *scope creep* that can threaten timelines, which (at least in enterprise software) are usually driven by product management or executive leadership based on a variety of market factors that might not be apparent to development. Even in organizations that have a mature project management function, it is incumbent upon product management to *keep committed timelines front and center* and to help your engineers remain focused on the most important parts of the problem such that features ship on time and do not continue in perpetuity until they are "perfect."

Have your team's back at all times
As you talk to your development counterparts, make sure they know that you win and lose as a team and that, if necessary, you will accept blame if something goes wrong. Great development practices can often involve taking significant risks—betting on an unproven but promising DBMS or

implementing a JavaScript framework that they think will scale, for example—and this is often where the best innovation happens. Never blame development for earnestly trying to do what you've asked, even if they failed. This alone will win you many acolytes among your developer colleagues. Never use the word "they" when you can use the word "we," instead.

Defer credit

Similarly, never use the word "I" when you can use the word "we," instead. Most of the time, despite being the ones who spend long days for weeks and months designing and perfecting a solution, engineers aren't the ones on stage at conferences or presenting in company-wide meetings on the things they've done: you are. It can be easy to forget who made those opportunities possible. In many cases, development is like art: the creator of the solution is inherently proud of it, and feeling acknowledged and appreciated means more than just about anything else you can do. When you communicate with your engineers, and with others about your engineers, make sure to acknowledge their efforts whenever and wherever you can.

Don't unduly alarm them with tales of escalation and attrition—but don't lie, either

As their connection to the customer base, your attitude and outlook on your product's future carries a tremendous amount of weight with development. A product manager who is all doom and gloom inadvertently conveys a sense of fear or discouragement to the development team. Be honest when customers are unhappy, but be optimistic—this is the team that can start to turn things around. Whenever you hear a CEO report missed earnings but convey confidence that things are going great, think of how you can similarly communicate bad news in a positive light to your engineers. But if bad news doesn't need to be communicated, skip it.

Wherever possible, help them see how what they are working on is going to solve real customer problems and make your company lots of money. Everyone likes to feel that their contribution to the company matters. Share stories of deals won because of something that one of your development teams did. Talk in detail about the customer feedback that you're using to justify asking them to do a project for you. Tell them how you are going to convey to the sales team what they are doing, and how you expect it to benefit the sales process.

Design

Conventional wisdom states that enterprise software often trades elegant user experience (UX) for customization and flexibility. Although that might have been true historically, the rising generation of enterprise software users (like us) grew up with the iPhone and Instagram; we are digital natives, and we expect better. Not every enterprise product manager will interface frequently with your design (which we are using as shorthand for "UX design") team, but in many software organizations the relationship between product management and design is what makes the difference between an average UX and the kind that delights users.

In fact, what you might experience is that one of your chief responsibilities is bringing development and design together to brainstorm and hash out ideas. For example, you might come back from a meeting with a customer excited about a point of feedback; the customer has clearly articulated a use case and you believe solving it would be in line with your product vision, useful to everyone, and achievable. The next thing you might do is get an engineer and a designer into a room with a whiteboard to share what you learned. If you're lucky, articulating the use case (i.e., the customer problem) might be the only thing of substance that you add to that conversation; from there, the light bulbs begin going off in a good designer's mind as an experience comes into view for him.

Much more than just a creator of wireframes and mock-ups, your designer cares about how the user feels while engaging with the product; he considers how to solve the customer problem in the most elegant, holistic way possible. This is evident as your designer talks passionately about how a software product in a completely unrelated market solved a similar use case. Then, engineering chimes in with thoughts about how the designer's proposal might work (or not!), and you're off to the races as that exchange produces one outcome after another. It can be exhilarating to see how ideas become designs and designs become product! It usually doesn't go as smoothly as we've laid out here, but this is one ideal to aim for.

Product management, design, and development can be thought of as a triumvirate that makes great products happen. Product management supplies a thorough understanding of the customer problem and the market opportunity; design discerns the user problem and a deep knowledge of human-computer interaction patterns; development adds the practical aspects of software creation (like resource constraints, technical limitations, etc.). Without the customer problem and market opportunity, design and development can lose focus on the product vision and build the wrong features. Without the user problem and knowledge of how humans engage with software, the problem might be solved in an inelegant or even unusable way. Without the pragmatism of engineering, obviously, software would never actually ship. All three roles are critical to making great software.

In fact, given that enterprise software is not traditionally known for being easy to implement and use, when properly involved in your process, your design team can become quite a powerful competitive advantage as they look for ways to solve new and existing customer problems in simpler ways. If you have the opportunity to work closely with design, encourage them to always be on the lookout for opportunities to streamline and reduce, focusing on putting the functionality that is most important to solving the problem first, and removing the clutter that is too often endemic to enterprise software. Even in this segment, products win despite not having as many bells and whistles as a competitor because they are a delight to implement and use. And if your product has all the features *and* good user workflows, it will be tough to beat indeed. Design can help you get there.

HOW TO INVOLVE DESIGN

The first and most important rule of working with your design team is to make sure that they know what projects are coming down the pike. In other words, help them get familiar with the product strategy and roadmap. There will always be this or that unexpected small project that requires some input from the design team to make it successful, but if your design team has a good sense of what you are hoping to ship and when you will need design input, it can prepare. Good design takes time. Your designers will want to do their own research: meeting with users, exploring tools that have solved similar problems in creative ways, developing personas, and more. When you spring a project on the design team at the last minute, this process becomes rushed, and your design team will miss important context.

Beyond simply allowing your design team an opportunity to prepare and research a user problem, you will hopefully find that keeping your designers in the loop provides you with a group of partners who can help you innovate. Although they might not have time or the mandate to become market experts in the same way that you do, good designers are some of the most creative thinkers in your organization. Your market expertise plus their focus on users and workflows can produce some of the most impactful innovation of your career.

Ben experienced this as part of the Adobe Analytics product management team in 2012. Adobe was looking to extend its user base from the digital analyst to the less data-focused marketer. The customer problem—the growing need for more roles in the enterprise to use data and analytics—had been clearly defined, but it was the design team, not the product team, that created the UX paradigm of "data foraging" for the analyst, and curation and data consumption for the marketer, which resulted in the highly successful Analysis Workspace tool. Product management and engineering would not have solved this problem the same way, and if design had simply been presented with a proposed solution late in the process and told to make it look nice, the result would not have been nearly as impactful to Adobe or its customers.

One of the other essential benefits of involving the design team in the product development process is that it gives you an opportunity to provide feedback and iteratively hone a design before the product is built. We recommend holding both regular and ad hoc sessions with your design team to review its work. In many cases you, as a product manager, will have market or customer insight that hadn't surfaced previously in the design process but that can help keep your product on the right track.

Lastly, use your designers to get feedback on designs and prototypes before your engineering team spends too many cycles building the wrong thing. Your engineering team might also take the lead on some user testing (both in the form of prototypes and alpha/beta tests), but your designers are the tip of the spear. Facilitate meetings with customers where your designers can explore what users like and dislike about a product design. Ensuring that users have a chance to give input at each stage of the development process is a great way both for your designers to hone their work early and for all involved to learn more about what your users' motivations and preferences are.

COMMUNICATING WITH DESIGNERS

For many enterprise product managers, design (i.e., UX design) is the only team with which you will work as closely as you do development. It is sometimes said that product management owns the *customer*, and design owns the *user*; if this is true, bridging the gap between how your product is *bought* and how it is *used* is one of the most important things that can come out of your communication with design. Here are some ways to make sure that communication channel stays open and clear:

Establish core tenets for product design and use those to stay aligned
　　Of course, you will want to help designers understand the business value of anything you are asking them to do (just like engineers), but designers are

sometimes prone to wandering away from the vision and business value as they encounter new and bigger ideas. Establishing the core tenets of the product or features they are designing in advance and returning to those often in your communication is key. When the design team drifts from those tenets, this makes it easy to get them back on track. As one senior designer at an enterprise software company put it:

> Help us understand the business value of a project, and keep reminding us of that value throughout the process. Also, you should help us in identifying the other few design principles that will be serve as the team's North Star. That set of agreed upon items serve as a great rallying point when you and your designer don't see eye to eye on something.

Let your designers play five years out, but be firm in bringing them back to the next six months

Many designers are visionaries, with brains that make connections between ideas in unique ways that lead to tremendous opportunities for innovation. This can sometimes lead them to take a very specific design request—"let's make it easier to install our software"—and turn it into a wholesale reinvention of your platform. Great things can happen when they are allowed to explore on a grand scale, but do not be afraid to hold them to deadlines for coming up with ideas to solve the problem immediately at hand. In your communication, remind them that you are all committed to releasing features to market in the short term, and allow them to explore both the short-term and long-term in parallel.

Try not to say that an idea is bad; instead, propose a better one

Earlier, we discussed how code can be like art. That is doubly true of UX design, which is often quite literally art. As such, you'll want to be careful dismissing ideas from your design team. One design manager told us of a project in his graduate program for which the professor made each student present their ideas and then had the other members of the class tear it to shreds as a way to "toughen up" these future designers and to teach them not to fall in love with their own work. Not many designers have had that kind of training, and constantly trashing their ideas can lead to long-term resentment. Couch your criticism in the form of improvement on the ideas that have been put forth.

Convey that you recognize that a good UX designer is more than just an Adobe Illustrator guru

No designer likes to be thought of as a mock-up monkey, producing beautiful screens with no context or clear purpose. Designing a true UX is a lot more than changing the color of a button. Concepts such as workflows, memory, personas, accessibility, intents, and problems-to-be-solved are the tools of the trade for true UX design. When you are giving feedback or brainstorming with your designers, make sure they know that you value these approaches to design. Talk about your requests and your customers' requests in these terms. Make these concepts a part of every conversation that you have as you work together to design the product so that your design partners can produce actual *experiences*.

Executives

We all work for someone. Whether you're a front-line product manager or managing the entire product group, eventually you're going to be asked to give an update, status, or summary report to executives. Depending on the size and product portfolio of your company, this might include C-level management or someone significantly downstream but still very important. Either way, you obviously want to prepare carefully for these opportunities.

Typically, unless there's a specific all-hands-on-deck issue at play, executives want some combination of the following:

- Status check: how are we doing?

- Roadmap: where are we headed?

- Strategy: are we doing the right things?

Fortunately, as the product manager, you're already on top of all of these things. In fact, the most difficult part about presentations like this is how to summarize the relevant knowledge into something that is consumable in the time and attention constraints you face. One way to do that is to put yourself in an executive's seat. Perhaps she is already an expert on all the minutiae of your product, but there's a good chance she's not. A good executive doesn't necessarily need to be; after all, that's your job. Particularly in larger companies, with lots of different product portfolios competing in tons of different markets, there's a good chance that executives will look to you to not only be the *product* guru, but the *market* expert, as well. (We talk a lot more about how to do that in Chapter 7.)

But for now, you have 20 minutes with a senior vice president to give her and her direct-reports a full report on the aforementioned three bullet points, any of which could easily be an hours-long discussion by itself. Here's how we might approach this:

Status check

Remember that at the end of the day, particularly in enterprise software, everyone is evaluated on sales growth, including your executives. How are sales faring against the annual plan? Think green/yellow/red-light guidance here, which you can back up with seller feedback.

Executives are often interested in user growth and engagement metrics, too, but have trouble putting them into context, particularly as they relate to sales performance. Back up your green/yellow/red-light guidance with this context. (By the way, we go into more detail on measurement in Chapter 5.)

Roadmap

A high-level quarterly roadmap should more than suffice here. Outline on the epic level, not individual stories. Bonus: it usually fits on one slide.

This is an area where breaking releases into themes is doubly effective, both as an organizational schema and also as a way to "bucket" all the cool stuff you're working on in a particular period. It's easier for audience members to wrap their head around themes like "mobile security enhancements" or "database infrastructure" than you droning on about one-off stories like HDFS cluster upgrades.

Strategy

You'll have to forgive us just a little snark here: every executive thinks they know business strategy really well, regardless of whether their hunches are correct or even applicable in your context. Needless to say, this is not a fruitful debate to engage in during your meeting. It's smart to lay out your product strategy in the context of your executive's broader vision.

Overall organizational strategy is not typically something in which front-line enterprise product managers have much say. More likely, your best course here is to demonstrate ways in which your roadmap hews to your organizational strategy (and the corporate vision discussed in the last chapter), and to provide data points that support (or, if necessary, counter) the current strategy.

There is a "Goldilocks zone" of strategic planning here. Many companies "over-strategize" their execution, leaving themselves little room to maneuver given new circumstances or information; others fail to strategize properly at all. Try to read which situation you're in and adapt.

At a high level, executives, like anyone else, look to the product manager to know "what's going on" around your product. If there are issues facing your platform or product, you should be in the informational loop; if there's a customer who's hopping mad (or who loves you!), you're on deck; if someone has a question about something on the roadmap, you're the person who owns the answer. Your answers don't always need to be happy ones, and if you don't know, that's okay, too. Particularly with executives, the company relies on the product manager to be the "truth teller" of the organization as well as the executional lead who clears issues from the docket. If there are caution or warning flags to be raised, the product manager needs to do it. That can be uncomfortable, because it puts you on the spot; but, well...that's the gig.

A good practice to get into when communicating up the ranks about your product is creating a regular, standardized status update document. This can be something as simple as a few slides (if your company uses those), or even just a formal document. It should include updated, summarized quarterly data on the major metrics you follow, like sales and closed leads, average deal sizes and germane seller feedback, roadmap goals in the quarter (or whatever other time interval) to come, and maybe a line or two about relevant news from your market or competitors. Try sending this up the chain and around your organization. You're likely to find that many people really appreciate the summarized information.

COMMUNICATING WITH EXECUTIVES

There have been entire books written on the topic of executive communication, and we won't rehash the topic in its entirety here. Still, as a product manager you will likely have above-average exposure to senior leadership compared to many of your colleagues. Often you will need executive buy-in to ensure that your projects are funded by development and marketing, and in other cases you might need to convince executive leaders to buy a company, to consider a new market, or rearrange priorities—all recommendations or requests that can originate in product management. Because the stakes can be high with an executive audience, your communication skills need to shine:

Tie everything to the corporate vision and the product vision

The single best way to help your executive audience connect with your ideas when you communicate is to tie them back to the company vision (and, within the company vision, your product vision). This is more than just keeping your initial communication high level; it means not leaving a shred of doubt how what you are discussing connects to what your company is trying to do. You can leave it to your executives to make that connection themselves, but from where they sit, 50,000 feet above the fray, it might not be as obvious to them as it is to you. Restating your company's vision and then drilling into whatever topic you intend to discuss is a great way to bring focus to any presentation, phone call, or email, and also shows that you have internalized what your executive team has put forth.

Everything begins with the business metrics

Nobody feels successes and failures more acutely than your executive team; to put it crudely, they have a lot more to gain and to lose as the business grows and shrinks. Relating your communication to the metrics that matter most—Annual Recurring Revenue (ARR), bookings, retention, and customer experience—helps put your ideas in context and show executives exactly how your projects are influencing the world that they care about. This won't always be possible in every type of executive communication, but whenever possible, try to relate things to what your leaders care about the most. For example, consider weekly updates that your executive(s) might ask you to provide.

Although many thought leaders will tell you that a "good product manager isn't busy," truth does not always match the ideal. Your calendar might end up packed full of meetings with various teams about diverse topics and projects. It can be tempting to say things like, "Met with development team about Project Phoenix and came to understanding of path forward." But this doesn't tell your leadership why they should care what you're doing. Avoid such travelogues. A better update might say, "Project Phoenix is on track to deliver a 30% lift in cross-sell six months after release. This week the team made the following decisions to help us reach that goal," followed by enumerating how you're going to get the product to that objective. It isn't that your executives don't care who you are meeting with or what you are accomplishing; however, expressing your activity in terms of its impact on the business is a better way to convey the value you (and product management) are adding to the company.

Bring data and tell a story

As a product manager, your in-depth knowledge of customer problems is unique in your organization. Conveying that understanding effectively to executives can produce meaningful change and get innovative ideas prioritized. You cannot assume that your senior leadership understands customer problems the way you do. The best way to help them grasp the frustration that customers (and the market) feel is to tell a story that blends personas and data into a simple but meaningful narrative. For example, you might begin by telling the story of one of the individuals you have met in your customer interviews and how this individual is currently prevented from growing her business. Relating that to market data ("This problem isn't limited to XYZ Corp; a survey of CIOs by Forrester shows that 73% of organizations rate this problem as one of their most troubling...) creates a strong blend of emotional resonance (with the end user's frustration) and logical reasoning (supplied by the quantitative data). This is actually a great tactic when communicating with nearly anyone, but with the limited time you have in front of senior leadership, persuasion in a few sentences or slides is key.

Complain about resourcing with caution

One of the most common product management pitfalls is to decry a lack of investment in development or design as the reason that a critical project can't be completed. Resource constraints are real, but to the team managing your company, complaints like these can come across as defeatist and pessimistic. A better way to convey that you won't be able to deliver on what has been requested is to explain what you *can* do, and why you have chosen to prioritize the projects that you have. That said, if critical projects are not receiving adequate resources, it is the product manager's job to raise a flag early and make the case for aligning resource allocation with the organization's strategy. To the extent there's a gap between the two, identify it early.

Enterprise software companies are often large, matrixed organizations, usually by necessity. As so often is the case, relationships and an understanding of the internal "social graph" are invaluable tools to get anything done and to route the right tasks to the right people. To the extent that you can navigate your organization more effectively, you will quickly become a much better leader and advocate for your customers and the product. So, get out from behind your desk and meet some people!

Marketing

> *"[Marketing] is the whole business, seen from the customer's point of view."*

> **—PETER DRUCKER**

Great marketing is like a jetpack for your product. Software companies that are able to figure out the optimal blend of brand, solution, and product marketing and use it for demand generation connected to a capable sales funnel set themselves up for enormous success. Smart marketers are able to analytically demonstrate their "MROI" (marketing return on investment), which allows executives to approve bigger budgets, which in turn leads to more revenue—a beautiful pinwheel. We've seen marketing organizations able to show an MROI of 5, 10, and even 15 times, whose biggest problem was keeping up with all of their profitable growth.

Unfortunately, we all know that great marketing campaigns, like great marketers behind them, are rare. Much more common is middling, unclear, uncompelling, timid, or outright bad marketing. (We could tell stories!) Bad marketing is more like wearing a jetpack that doesn't fly: it's just heavy, cumbersome, and occasionally explodes in your face.

To reach its full potential, marketing relies on product management for guidance. The full gamut of product management's interface with marketing (usually product marketing, creating an awfully confusing battle of "PM" acronyms) is impossible to outline here, but suffice to say that a weekly, or twice-monthly at least, check-in is a best practice. We most often find ourselves working with marketing on three themes. Let's take a look at them.

PRODUCT

This is an obvious one. product management should be marketing's go-to contact for expertise on the product, features, benefits, and value proposition. When there is an update or new release in the pipeline, marketing should be brainstorming with product management long ahead of time to ensure the relevant launch content is distilled and presented to the market in the best way possible. When a new product is being launched, product management needs to confirm that marketing grasps the relevant market details (the target user, industry, and value proposition) inside and out. In SaaS settings, product management, along with development, should also help marketing get set up with the appropriate demo and sandbox accounts. Everyone in Marketing should be encouraged to have a login into a sandbox environment where they can get a feel for the product and not break anything.

CUSTOMERS

Product marketing (and sales) should be the only functions that rival product management itself for a deep understanding of their customers and end users. As product management, we need to compare notes with marketing on who our customers and end users are, what their jobs are like, and how our product makes that job easier. Sometimes, we can teach marketing a thing or two, but as often as not, they teach us! Only with full visibility of the customer can marketing do its job of crafting stories and narratives around our product that connect with customers and users alike, and both teams need to constantly reevaluate if our assumptions about them are correct.

MESSAGING

As product management, we own the product vision, embodied in artifacts like the roadmap. But turning that vision into consumable messages that resonate with customers is literally a definition of marketing. Coming up with good messaging and delivering it well are both difficult, incredibly underappreciated skills. (If you don't believe us, go look at nearly any two examples of technical messaging, and try to tell them apart.) Let marketing handle this; they're the experts. Product management should focus on giving directional input to ensure that the messaging aligns with what the product does, its business value for your target industries, and where it's headed (the vision). Regularly revisit this and make sure marketing's messages remain consistent with yours.

MANAGING CONTENT

One of the most common requests of marketing by the larger organization is for content: customer-facing, market-facing, and internal. On most external collateral, marketing should be good to go—communicating with customers is their specialty, after all, and if anything, they'll need at most directional guidance from product management. Internal content, however, can be a different story. Internal content can range from sales enablement materials and pricing guides to competitive comparisons, ROI business cases, and "battlecards" (i.e., one-page summaries of solution details and key benefits) for sellers in competitive sales situations. (See the end of this chapter for an example battlecard.) Product management tends to have more involvement with these. Wherever possible, we recommend tasking Marketing with taking the lead on developing tools like these, with product management providing input as necessary, and for owning their ongoing development (i.e., version control).

Along with owning customer-facing content, one thing we'll touch on in the next section about sales is how to get sellers the content they need. In our experience, very few companies have a great internal content management system for this. In small organizations, it often boils down to something along the lines of, "Just email Bill; he's got the latest and greatest." This is, of course, the easiest option for a single busy seller, but it obviously doesn't scale. In organizations with dozens, hundreds, or thousands of content consumers you need to reach, a superior solution is a central online hub where everyone can self-serve what they need. (Many of us shudder just thinking about the SharePoint sites we've all wrestled with to establish this very "single source of truth.") Even when that sort of system is in place, invariably you will still receive panicked emails from sellers and others asking you to just send them what they want. We'll go into much more detail about how to address this in the next section on sales, but for now, if this describes life in your company, don't worry. It's not just you or your company. Keeping all of the marketing collateral needed for a multiproduct solution organized, findable, and current can easily be one person's entire job, and it is often a thankless one.

For the sake of our marketing friends' sanity (and our workloads), clearly delineating roles and responsibilities around content creation (and maintenance) is crucial. There should be no question of whose responsibility it is, say, to keep the sales decks updated and to drive training content for sellers. Sometimes, this can be as simple as having a talk with marketing and making a decision; other times, it's a good topic for your regular check-in, for which one item on your agenda should be looking ahead at events, trainings, or the like for content needs.

What If I Don't Have Product Marketing Support?

It isn't uncommon for certain products, or even product areas, to lack dedicated product marketing support. This can happen in smaller organizations because of a lack of headcount, or in very mature ones, for any number of reasons (including...lack of headcount). In these cases, what is product management to do?

For these circumstances, there is really no other option but to brew some coffee, fire up PowerPoint, and create a "sales kit." One downside to product management is that we often play "cleanup" for other functions in a pinch. In our opinion, the absolute minimum your enterprise product's arsenal of marketing collateral should include is the following:

- A basic sales deck that explains what the product is, what it does, its key benefits, and messaging. This usually includes screenshots as well, and plenty of swappable slides that sellers can use to tailor a deck to a prospect.

- An internal-only sales enablement deck, which is mostly the same as the aforementioned customer-facing one, with more technical detail slides at an appropriate depth, pricing and ordering information, internal contacts to ask for more information, and lots of swappable screenshots.

- One-page guide (pro tip: landscape mode!) with key product value propositions, messaging, feature details, and competitive points. (See Figure 4-1 for an example of a battlecard.)

- Optional: official product data sheets, whitepapers, and links to official marketing blog posts or relevant recent webinars

---

One final word: if you're lucky, in your regular rounds with marketing, you'll notice that certain gung-ho marketers are the ones who really become adept at your product. They'll be the ones who can run their own demos, and run them well; who know the documentation and can go look it up without guidance; who, over time, develop a reputation with sellers and the broader organization for being the "go-to" source for information. These folks will become an outsized resource for you to lean on and are outstanding candidates for future product management recruits. Keep an eye out!

Gopher Content Management - Killer App Inc.
www.CompanyURL.com/GoGopher

Sales Battlecard: Gopher CMS

Solution Overview: Gopher CMS allows one-click access to a company's full content library, with seamless integration into many 3rd-party applications. With embedded analytics and rich editing features, Gopher CMS enhances content marketers' capabilities while making content management simpler than ever.

Target Audience: Heavy content consumers
- Content & Product Marketers
- Salespeople / Account Executives
- Knowledge management pros

Top Customer Pains:
- Sales & marketing content is hard to find, access and edit
- Managing & requesting user/user group permissions is difficult and confusing
- Content creators do not know what content is highly viewed or useful
- Content is hard to embed with other applications, like email or messaging

Key Capabilities:
- QuickSearch feature offers instant library-wide location of needed documents
- QuickView feature allows content creators to glance at document without a full open
- DocTrack feature provides rich analytics, including document opens, pages viewed, shares and more
- Content syndication to downstream applications is available in one click

Differentiators:
- DocTrack provides superior insight
- Cloud storage makes content easily available to teams anywhere
- OAuth & SAML support for enterprise authentication flexibility

Customer References:
- Big Insurance Inc. shortened quarterly sales cycles by 38%
- Catalog Retailer improved product targeting, creating 250% ROI
- SoftwareCo grew sales efficiency by 355%

Pricing/Contract Info:

50 users	500 docs	$1000/month
250 users	5,000 docs	$2500/month
1,000 users	20,000 docs	$5000/month
5,000+ users	Contact Sales Mgmt	Contact Sales Mgmt

Discounting available between tiers.

Additional Info:
- Technical datasheet
- Integration connectors
- Pitch Deck

Internal Contacts:
Product Marketing: John Doe
Product Management: Jane Doe
Partner Sales: Joe Blow

Figure 4-1. Example sales battlecard

COMMUNICATING WITH MARKETING

Marketing, particularly product marketing, is product management's right-hand partner in interfacing with the market. As such, communicating effectively with them should become natural as you practice it. When communicating with your marketing colleagues, keep these principles in mind:

Use product marketing to help turn your message into something perfect for your sales teams

> Sometimes, it is difficult for product managers to separate themselves from the products and features that they are working on and to be able to convey the value succinctly. Sales comes to product management looking for a few sound bites that will make their prospects salivate, and they get back what sound like technical specifications. Your communication with product marketing is an opportunity for you to "try out" your own translation of technical specs to market value, with an audience that won't immediately go try to sell it five minutes later. Product marketing should be adept at helping with that translation, but get them involved early and communicate changes and thoughts often to ensure that what gets in front of sales is as polished and convincing as possible.

They are often the ones selling your feature or your vision in the market, so don't share more warts than necessary

> When it comes to selling your vision at industry conferences, when speaking with industry analysts, on your website, and with the media, nobody is more potent than product marketing. These forums are critical to the success of your product, even if they do not directly result in new leads for your sales team. The nature of these public forums is that you will want to put a positive spin on your product capabilities and value propositions. Thus, product marketing doesn't always need to know every limitation of a new feature that you are building. This isn't helping them to lie, but it is allowing them to tell the most positive story possible without being burdened by more detail than they need.

Be technical, but not too technical

> Whereas product managers might come from a more technical background or have a minimal aptitude for technical topics, product marketers rarely do. Explaining how your product works is great; a product marketer who can speak intelligently about the capabilities of your product *and* the

higher-level benefits and value propositions of your product will take you far. But, as with the earlier point, product marketing does not need to be inundated with the minutiae behind the product. Helping them understand what the product does allows them to explain how a customer can use it to achieve some objective, and helps them craft stories that sales can use to win deals; helping them understand the trade-offs that the development team made between three different JavaScript frameworks, however, is usually more detail than they need or can synthesize into a story.

Above all, help them tell the story

Working with product marketing can be like working with sales, except that product marketing does not have the same intense focus on the current quarter of business, and working with product marketing scales your work with sales in the sense that product marketing can be a bullhorn for you, sending news of your product far and wide for your sales team to sell. Like sales, product marketing might need some extra help and context to understand your key value propositions and success stories from customers. The best thing you can do as you communicate with product marketing is to give them the raw pieces of a great story. If you have a good counterpart in product marketing, that person will be among the best in the world at turning a customer problem, a value proposition, details about a solution to the problem, and customer examples into a story that sales can use all over the world to win business for you. Make sure you are always giving them those raw pieces and then helping to fill in whatever gaps are necessary.

Analyst relations

We couldn't very well ignore this minor, but important, stakeholder group. Analyst relations doesn't always fit organizationally under marketing, but its aims are pretty well aligned, so this is as good a place as any to address this function.

Industry analyst firms like Gartner, Forrester, and any number of boutiques are a mainstay of enterprise software. The startup crowd loves to poke fun at them (often for good reason) because they're culturally and strategically aligned with and for big companies. There is also criticism, which is somewhat unfair, that analyst research is tainted by a "pay to play" bias: vendor firms pay the analysts, who are also the ones judging the vendor firms. On the other hand, you might be surprised to hear the percentage of brands in your target market that rely on these analyst firms to help sift the viable software solutions from the ones making noise in the market without delivering real value. As a product manager in enterprise software, knowing how to use analyst relations to your benefit is a smart skill to have under your belt, but it requires clearly understanding the delicate give-and-take required.

Analysts don't have a crystal ball. They mostly look at the same reports you do, and obviously have the benefit of far less sensitive internal data. That said, they do talk to a lot more of your competitors' customers than you do, and probably a fair number of your customers, as well. They use those interviews to build market narratives that go into their research. Most of us at big companies have access to key analyst research on our markets, which should naturally go into the big funnel of your information diet.

The entire reason why enterprise software companies invest time and resources with analyst firms is because our customer base reads their research, and customer shortlists for vendor selection are often influenced by the most recent Gartner Magic Quadrant or Forrester Wave. Less frequently, companies will actually engage with an analyst to help narrow down their shortlists or even perform vendor selection based on a company's contextual factors. For all of these reasons, it's worth setting aside a day every quarter or two to have an Analyst Day with the lead analyst on your market from a given firm. Ask them to come and present their latest research, bring questions you have about customers in your market, ask for feedback on your *high-level* product roadmaps, and ask for the analyst's views on how you're doing. We recommend getting input from more than one firm (and, of course, it need not be just Gartner or Forrester; these are just well-known examples). Analyst retainers might be useful for you, or they might not be. We personally run into few situations in which we feel the need to call up an analyst about a particular problem on any given day, but your mileage may vary.

In doing all of this, be aware of how analysts' incentives can affect their research. Analyst thought leadership is both their product as well as their marketing. It is not uncommon to see prominent analysts make dramatic, provocative, or even outlandish predictions, because the penalties for being wrong often aren't really that high. Moreover, analysts, like everyone else, are heavily influenced by marketing. Their perspectives on market drivers, influences and trends, what defines "market leadership," and so forth rarely venture very far from the general consensus. Analysts are much more likely to fluently speak the language of your market, and again, they talk with a lot of companies—but as humans, when we're personally invested in a narrative, we tend to process new information in a way that supports that narrative. We recommend taking analyst input seriously, but not as revealed truth handed down from on high.

And not to repeat ourselves, but of course you should endeavor to know your market as well as they do, if not better. Be your own analyst first.

Sales

> *"The sales department isn't the whole company, but the whole company better be the sales department."*
>
> **—PHILIP KOTLER**

Sales is a surprisingly unsexy topic for a lot of product people. There are, of course, volumes written already about sales, how to sell, effective persuasion, and the like, but most product management literature hardly touches on the subject. We think this is because most of that literature is focused on consumer businesses, where it's rare for product managers below very senior levels to interact with a sales team (that might be selling advertising or some other derivative, not the product itself). In that world, it's almost as if "shipping product" is where product management's job ends! Naturally, those of us in enterprise software know differently. "Shipping" is only the start. Now, not only does someone have to buy it, but odds are good that the customer is going to buy from an actual carbon-based salesperson.

Remember: we're all in sales. There are arguably only two real roles in any enterprise software business: engineering and sales (though, ultimately, even the engineers are in sales). Using an appropriately broad definition of the word "sell," it is everyone's job, at all levels of the company, to "sell" the product, whether they are giving product pitches or not. Be careful about hiring anyone who doesn't embrace that. Part of what makes product management such a cool gig is that we have a foot in the worlds of both engineering and sales, and having a passion for each is an important part of conceptualizing our role.

Understanding enterprise sales is incredibly important to being an effective product manager. The reasons are pretty obvious: someone must buy what we make, or there is not much point in making it. Enterprise software sales are, have always been, and will likely always be pretty complex. Our products are sophisticated and specialized, and their value propositions are finely tuned to the specific problems people have in a given industry. It's often a long and twisty road between a prospect's first signal of interest to a person actually using our product. The process often looks something like this:

1. Resourcing discussions happen internally at the customer.

2. Budgets (for both software and sometimes headcount) are allocated.

3. Prospect interest is signaled. Sales calls are made.

4. More sales meetings. Demos are done. Designing and solutions happen.

5. Negotiations on price, components, architecture, and so on.

6. Contracts are signed. Everybody is happy.

7. Implementation process. This can take as little as a few weeks to, in some cases, a year or more. (Do not underestimate the implementation process in enterprise software. Rolling out a $50 million purchase to tens or even hundreds of thousands of employees can be brutal.)

8. Solution training (often simultaneous with #7).

This is a highly simplified flow, but you get the idea. Enterprise SaaS sales cycles are routinely between 6 and 12 months. This can be a big advantage in some ways, like when you want to start telegraphing capabilities to new prospects that are still two quarters away on your roadmap. In other ways it's a drag, because longer sales cycles make it harder for prospective customers to pay you and use your product.

As an enterprise product manager, you should understand *exactly* how your product is sold, and what the hand-off process is like, from the first prospect a lead development representative (LDR) follows up on, to the seller who takes a qualified lead and sees it through to a sale. The people at each step of that sales process need different things from product management and can feed different kinds of valuable information back to you.

THE SALES ORGANIZATION

Sales organizations come in all shapes and sizes. Assuming that your company does business internationally, you will almost definitely have sales teams based in other countries that are focused on customers and prospects in that area of the world. If your company offers multiple enterprise software products or portfolios, you might also have sales representatives who specialize in one product and who will join a conversation with a prospect to help sell their product. With larger software firms, you might even have "sales reps" whose job involves very little actual "selling," as you might define that word, but instead cultivate relationships with decision makers and locate budgets within a prospect's organization so that they can bring in other sales reps to do the actual selling. You will get the lay of the land in your own organization as you work with sales teams on deals.

At a high level, however, we can categorize our sales colleagues into three groups which are near-universal at every enterprise software firm we have seen. Their titles may vary, but the roles are consistent.

Account executives own the relationship with the customer or prospect and are typically the ones who bear the most pressure in terms of quota (the amount of ARR that leadership has assigned them to generate). Their most important job is to manage that relationship and try to match a prospect's business need with the solution the account executive is selling and seeing it all the way through to contract signing. Typically, their knowledge of your product will be limited to high-level value propositions and basic descriptions of various features, but they should know how to talk about the benefits you provide and how the software can solve whatever problem the customer is facing. Usually your counterparts in product marketing will take the lead on training and enabling account executives to help them know what to say about each new product release.

Note

When we say "sales rep" in this book, we are referring to account executives.

Sales engineers (often called *solution consultants*) partner with account executives as the more technically minded arm of any discussion with a prospect. The good ones are adept at demonstrating the best differentiating features of your product and applying those features to the customer's requirements to find product–customer fit. They might not know the entire product as well as you do, but some of them will come mighty close. Because the solution consultants are more hands-on with your product, we find that product managers are better equipped to train and enable this role so that they are able to demonstrate and speak to the ins and outs of your product as effectively as possible.

Pre-sales/inside sales/lead development representatives are the front-line soldiers in your sales funnel. These folks do exactly what their title suggests: they develop warm and qualified leads to pass off to the seller(s) with whom they work. They are often the first point of contact for prospects replying to an email campaign, live chat, or phone call, and they spend a lot of time making *cold calls*. If you've never done that sort of work, take a moment and extend some compassion. These folks put up with a lot of rejection, but are also the source of a lot of your sales leads. They quickly learn about the tactics, campaign copy, and value messages that drum up leads.

A final key constituency to keep an eye out for here is sales operations. This can be a small team sitting somewhere, or it could effectively consist of a number of sales managers working in tandem, or it could even be just one lonely analyst squirreled away all by his lonesome. Whoever it is, in most high-functioning sales departments, particularly in larger companies, there is someone, somewhere working to consolidate and analyze sales data from across your company/ group. (See the sidebar that follows on pipeline reporting.) This analysis will often include basic customer relationship management (CRM) data, including number of leads, days between touches by sales, deal qualification stage, days between each stage, deal sizes, and hopefully even deal quality. The reason you care is because this information is a peek into each customer's disposition around your product, and invaluable intelligence about the market: did they happily pay a lot for it, or did they negotiate hard? Are they a "good" customer to have, or are they likely to walk at the slightest provocation? Knowing who is paying what and how they perceive the value for that price is inimitable and rare market intelligence to have.

Of course, there are plenty of others involved with selling your software, but most of your direct interactions related to turning a prospect into a customer will be with some form of these three roles.

HOW PRODUCT MANAGERS ENGAGE WITH SALES

Sales requests come in all shapes and sizes but often boil down to one of the following:

Presenting the roadmap and/or product strategy

In both a sales situation and a renewal situation, dangling the roadmap in front of buyers can be a legitimately effective way to get them excited about partnering up with your company for the next year or three. In enterprise software, prospects are often buying into the vision as much as they are into the current offering. Knowing that you are working to address *their* problems keeps them optimistic and gives them some air cover to go with your solution, even if it doesn't do everything they need right now. In some companies, product management owns the roadmap and keeps it very tightly protected; in others, the roadmap is made more-or-less publicly available. In both cases, the product manager can shed light on the strategy and roadmap that nobody else can.

Answering technical questions

Sales engineers can handle many of these, but there are plenty of cases in which really hairy topics go beyond their expertise, and someone who is more plugged in to the inner workings of the software is required to speak to a requirement.

Demos

Ideally, sales engineers should be able to handle product demonstrations in any sales situation. (It is, however, up to you to ensure that the sales engineers are trained on how to do the demonstration.) It is appropriate to remind the account executive of this fact and to offer to prep the sales engineers on the deal prior to the meeting with the customer if they need a refresher. The exception to this rule is functionality that has not been released yet, and on which the sales engineers have not been trained.

Sharing best practices

Sometimes, buyers need advice from someone who has seen thousands of implementations across entire industries. They have questions about governance, security, hiring, and team structure, or implementation. In reality, product management is sometimes not the best role to speak to best practices, because product managers often go wide but not very deep with any given customer. If you have a consulting function in your company, we advise bringing them in here if possible because they tend to go deep (but usually less wide).

As you can see, some of these requests are truly best serviced by a product manager, whereas others are not. Because product management covers so many disciplines, and is so connected to nearly every other team at the company, sales reps will often start with you. Do not be afraid to redirect them if there is another role that is truly better positioned to help win a deal.

When to get involved

As we mentioned earlier, working with sales in the right manner can be a great way to help you both reach your goals. Product management, as a more technical, more strategically inclined role (i.e., not on commission and not totally focused on hitting a quarterly quota) has credibility with buyers and users that sales never will, and so engaging in a sales conversation can give the situation the push it needs to get across the finish line. As a product manager, especially after you have been around for a little while and have been part of shipping valuable product features, your presence in a sales deal—taking time out of shipping product to meet with the prospect and discuss the product as someone who has a hand in defining its future—shows the prospect how much she matters. Becoming a respected, reliable resource for your field (i.e., sales) teams can also do wonders for your career, because there is so much executive visibility on certain sales deals. On the other hand, some sales reps aren't concerned about the other demands on your time, and will try to use you to do things that should be handled by others, including themselves. This is why you cannot (and must not!) become a sales robot, doing whatever is asked of you without discernment.

Occasionally we hear about a product manager who has responded to a request from sales to do a product overview (which any account executive should be able to handle) or to cover some other aspect of the sales process that should be textbook stuff for any sales rep by saying, "Sure, I'll do it. Just pay me your commission when the deal closes." We might laugh, but we can assure you that your sales rep who hears this will not find it funny (and we do not advocate actually playing that card!). Ultimately, you need to be judicious about how, when, and where you become involved with sales. We talk later about how you can gauge when to say yes.

There are essentially two kinds of engagements with your sales teams: a *push* interaction and a *pull* interaction.

We can call it a "push" interaction when you proactively reach out to a sales rep to offer your assistance in getting a prospect signed or a contract renewed. A good practice if you have pipeline data (see the sidebar that follows) is to find the most strategic (which might or might not mean the largest) deals each quarter and offer to personally sponsor them. This is great for a few reasons:

- First, it preempts any last-minute demands from the sales team. In many cases you will receive a frantic email from a sales rep who is already on site with a prospect and needs someone to walk them through the product roadmap 30 minutes later. If you step in to sponsor a deal, there is a much higher chance that you will have been in the loop from the start, which makes the entire company look much more organized.

- Second, although customer interaction is the currency of product management, prospects will teach you things that existing customers can't. They might be using a competitor's software currently, and offering to join demonstrations and Request for Proposal (RFP) calls can give you insight into how they evaluate your product versus the other guy's product.

- Third, selfishly, getting involved with the most strategic deals each quarter will make you and your team look tremendous in the eyes of executive leadership.

- Fourth, and perhaps most important, it allows you to ensure that you are spending your time on the most significant deals at any given time. By being proactive, you avoid spending precious time on a $6,000 deal while a $700,000 deal goes sideways.

This can begin with a simple phone call (you can email, but in our experience many account executives seem to prefer the phone) to Sue, the account executive on a large renewal with XYZ Corp, to say that you noticed her renewal is in the pipeline for this quarter, and that you would love to get involved to help make it a slam dunk. In our experience, you will receive one of two responses: either the deal will already be so well in hand that the sales rep won't see any reason to waste your time, or she will jump at the chance to have a product manager in the fold to help with this or that. You might still need to balance involvement in any given sales situation with everything else that you are uniquely positioned to do (market research, helping with product design, etc.), but if you've volunteered yourself in the first place, sales reps will understand that they cannot monopolize you.

Pull interactions, on the other hand, are when sales personnel are "pulling" you into becoming involved, and must be triaged differently. As we implied earlier, the tricky thing with pull interactions is that it can be difficult to orient yourself to the situation to ensure that your time is well spent. Here are some questions to ask first:

- Is it an existing customer who knows your product well and is preparing for renewal, or a new prospect who is considering a switch?

- How big is the potential opportunity, both now and in the future? (And is the sales rep exaggerating about the potential future account growth?)

- Is the prospect considering buying other products from your company, as well, or just your product?

- Have the account executive and sales engineer been doing their jobs, or are they looking to pass the buck?

These are all things you will need to ascertain when such a request comes in. A common maxim in product management is that the product manager needs to be good at saying no. This usually refers to denying feature requests that are not on strategy, but we can apply it just as well to working with your field team. There are going to be times when you simply need to say no. It is true that your team will be judged, at least in part, based on sales, but this does not mean that you can let top-notch execution of research, requirements gathering, design, prototyping, iteration, development, and shipping your product suffer in order to win a deal that could be won other ways. It is appropriate to remind a sales rep that you are responsible for the success of the *entire* customer base, and that there might be others who can step in and help get a deal done. In these cases, have sales look to product marketing, evangelists (also known as technical marketers), and others who might be able to address the request posed by the prospect or customer.

In addition to assisting our sales colleagues with their jobs, they have much to offer us, as well, in the form of invaluable market intelligence.

PROCESSING FEEDBACK FROM SALES

As we mentioned, many leads typically begin with LDRs (recall that these folks go by many different titles). Good LDRs, like good sellers, are worth their weight in gold to your business, but typically have little product-specific feedback for product management. It's good to be in touch with your LDR team(s) just to find out what they hear from the prospects they talk to. A lot of it will simply be noise —like the prospect once who really wanted to know if we could render the Tajik alphabet in our UI, and demanded screenshot proof. (Yeah, seriously.) But over time, you'll hopefully develop a sense for deciphering the nuggets of insight buried among the nonsense.

The next big sales constituency to consider are those solution consultants (aka sales engineers) again. These folks are a tremendous source of customer insight. Very frequently, they will meet with a prospect's technical, IT, or development team to talk through how your product's architecture will work with their existing systems—think, a lot of in-the-weeds discussions about data types, models and warehouses, databases, authentication schemes, permissioning, integrations, compatibility, and so on. As mentioned earlier, they also tend to give a heck of a lot of demonstrations. Strong solution consultants become experts at connecting your product's features to the prospect's specific needs and pain points, and they frequently come back with pointed advice on what messaging is working and what is not. Because they tend to meet with more end *users* of the product than executive *customers*, SCs also develop a keen sense for "user problems" that product management should consider, like ways to improve your UX or workflow.

Note

It's worth mentioning here that a lot of enterprise product managers we know considered doing sales engineering instead of product management. Sales engineering is a unique blend of product-level expertise, technical finesse, and business strategy. It also pays well. Just a thought.

Last, but perhaps most important, are the sellers themselves again. To reiterate: these men and women are the lifeblood of your business. Their work pays your salary. They take on a difficult, demanding job, a commission-based compensation structure and comparatively more risk than almost anyone else in the company, and we have to respect that. Great sellers are usually extremely sociable and likeable, but can also be a little pushy, aggressive, and very persistent, all of which makes them great sellers but which product management must take into account when evaluating their feedback.

Enterprise software sellers deal primarily with the executive stakeholders who own the buying decisions over your product. They don't typically care as much about "user problems" as much as they do "customer problems" (the business factors which influence check-writing executives' decisions over whether to sign on the dotted line). This often means that there is comparatively more "signal" in their feedback to product management. Product managers should take every opportunity to absorb this feedback, either structured (as in a monthly sales team meeting that you better be attending) or unstructured (go grab lunch with your West Coast sales lead when he's in town). That said, there are some important caveats to taking seller feedback:

- *Everything* is urgent to sellers. Sellers often feel like they will lose this deal or that one because the product is missing X, and if only you will commit to building X, they can wrap it up. Obviously, though, if the development cost of building X is significant, and it really appeals only to a small subset of customers, that trade-off might just not be worth it.

- When you participate in a large total addressable market (TAM), you might have hundreds or thousands of potential enterprise customers. A great seller is talking to a few dozen of those per quarter, tops. If you hear the same product wish-list items come up again and again, you might well have a pattern on your hands; or you might not. This is a judgment call, but consider whether what you're hearing is localized to a particular geography, industry, or type of customer.

- Frequently, seller feedback about the *product* turns out to really be more about the *messaging*. It's difficult for even great sellers to overcome crappy messaging about what a product does and the value it offers. A valuable exercise to ask your good sellers is how they would change that messaging or description of your value proposition.

- Understand seller incentives. It's really true: salespeople are "coin-operated," and properly so. Are your sellers incentivized to bring in bigger deals, or a high number of smaller ones? How does their commission work? Are there sales "spiffs" (i.e., immediate bonuses) or "escalators" in play? All of these will influence the kind of deals they try to land.

By the way, in many organizations, you'll notice that certain "celebrity sellers" begin to emerge. These are reps who crush their numbers every quarter, manage to consistently bring in big deals, get "President's Club" every year, and so forth. Particularly in growth-stage companies (somewhat less frequently in mature ones), these sellers get a lot of executive attention and their word carries great political weight. As product managers, we should get to know these sellers well. They are clearly onto something with how they communicate the product's value and often have very insightful views into the adjacent range of business problems that our product could address. That said, be careful of over-weighting this input. Star sellers tend to be put in high-value territories, which can skew perceptions and information. Politics within your organization can also privilege their input over that of others. This is something to be conscious about.

Great enterprise product managers stay close to each of these groups. You should be talking with someone from sales almost every day. When big deals come in or quarterly results are posted, product management should be one of the first groups to know.

Pipeline Reporting

While we're on the topic of sales, let's talk about pipeline reporting.

Pipeline reporting (e.g., a view of how many deals your company or group has in play, at what stage, for how long, etc.) is another extremely unsexy and underhyped but incredibly useful tool for product management. In our experience, very few companies do it well, especially at large, public companies where finance departments are leery of releasing too much sales data, even internally. Used properly, pipeline data can help product management not only understand what's being sold, how frequently, and where, but also in what sales stages prospects are deciding that the product just isn't for them. This is obviously a matter of great importance to marketing and sales, but if prospects are responding to your messaging in the form of inbound leads but then dropping out in large numbers the more they see and investigate your product, that has product implications that product management might need to address.

In general, sales or marketing should be encouraged to own a regular pipeline reporting cadence. Weekly, biweekly, or even monthly reporting might suffice. The pipeline should be broken down into reasonably granular territories for your business and show progression between your tracked sales stages for number of deals, deal sizes, and number of days spent in each stage. Major, outlier prospective deals should be listed out. Make sure everyone in product management gets this report, and make it an item in your regular meeting.

WAYS TO SERVE, AND INFLUENCE, SALES

To sell effectively, sales needs to be armed. Of course, they need a good product to sell—or, at least, not an egregiously awful one—but they also need information in the form of content. Typically, marketing is responsible for enabling sales with the content it needs (see the sidebar earlier in this chapter about the bare minimum elements of a product sales kit), but inevitably questions come up for product management. A good example is RFPs.

RFPs are basically long questionnaires put out by prospect companies looking for vendors to fill a need that the prospect company has. RFPs can often tell you a lot about what competitors a prospect has worked with in the past and what business needs they're most focused on addressing. Ideally, your sales team has already worked with the prospect and influenced their development of the RFP itself to be friendly toward your product; but if not, don't panic. RFPs tend to get extremely detailed (like the aforementioned Tajik alphabet request!) and can go off on some pretty weird tangents at times. Sales often needs direct product management assistance to come up with appropriate answers.

Here's a pro tip: grab copies of RFPs that product management is roped into filling out with sellers, anonymize it, and make the list of questions/answers available to the entire sales team. Boom! We've just saved you hours of tedious busywork and made your sellers' lives a little easier. Instant ROI!

Beyond RFPs, sellers frequently need to find and access an entire constellation of collateral about your product or portfolio. Often, the way marketing deals with this is to set up some form of internal knowledge hub or content management site where everyone can browse the latest and greatest material. In our experience, the complexity and difficulty of use of these systems tend to increase exponentially with the sophistication of the product, making it a huge headache for sellers to find the stuff they need. And as any enterprise product manager knows, the seller's motto is: "when in doubt, email the guy/gal in PM who usually has the deck I need." You want to avoid this situation if at all possible, less because it's annoying, and more because it doesn't scale and leads the people who *don't* know you yet to simply give up because they don't know who has that one killer business case.

One way we've found to avoid this situation is to keep up a regular line of communication open with your sales team. It can be as simple as a weekly email that you send to all sellers (or cascaded through their sales managers) with usable tips about your product, competitive news (manna from heaven for sellers) or interesting and relevant market news. In every single such message, you should include a link to that online hub of marketing collateral, and clearly point them to the stuff on your product. Sellers are inundated with email, but typically very good at figuring out what's useful (and what isn't) for building their pipeline. If they respond to this, that's good news.

One tactic that Blair used to employ was an end-of-week "Competitive Note" email to all sellers in his product portfolio group (a couple dozen). When there was intelligence or news on competitors to share, he included it; but more often, this was a medium for sharing "did you know?" tips on less-known features of the product, overviews of market trends, and short updates on what product management was working on. This note eventually became so popular and forwarded-on that it was widely shared around their company's division.

If you have an annual sales kickoff event, and/or regular sales boot camps in various geographies, make it a priority to attend as many in person as you can and to present refreshed material on your product. You don't need to go story-by-story; a quarterly breakdown of roadmap themes is usually fine, in addition to a retrospective of past milestones that you've achieved. Go through boot camp exercises with sellers, even if they're not about your product specifically. The best such exercises involve building business cases, ROI justifications, and explaining the value of various products in combination, or targeted for specific industries. Some companies bring in outside experts for coaching on messaging, overcoming objections, and win/loss analysis, which can be extremely effective (ask us for recommendations!). Particularly for multiproduct software platforms or portfolios like those we've worked on, exercises like these give you a vitally important go-to-market perspective that is never far from a good enterprise product manager's mind.

SUMMER CAMP FOR SALES: THE VENDOR CONFERENCE

There is one other type of sales interaction which is worthy of mention here because it can take up a surprising amount of a product manager's time: conferences.

In this case, we are not talking about conferences where you go as an attendee to learn about your industry and market, as described in Chapter 7. Here, we mean your company's own conferences: events like Adobe Summit, IBM InterConnect, Oracle Open World, and Dreamforce. Enterprise software vendors market these events as opportunities for your customers and prospects to come bask in the latest and greatest industry and technology trends, to network with peers, to learn best practices, and to have a great time. (Dreamforce has had artists such as U2 and Bruno Mars perform.) They usually last anywhere from two days to a week and include main-stage product announcements, demonstrations, and guest speakers, as well as smaller breakout sessions, often led by —you guessed it—product managers.

Not all enterprise software vendors hold conferences, especially during their early years in business, but if your enterprise software company already has a critical mass of customers, there is a pretty good chance that your company does something along these lines. We named four of the biggest annual enterprise software conferences in the previous paragraph; your company might hold smaller events, such as *summits* or *cafes* or *symposia* open to customers and prospects locally. Or you might hold a larger annual conference that isn't (yet!) on the same scale as those we just mentioned. Either way, if you are new to enterprise software, this aspect of generating business can sometimes be a bit overwhelming and disorienting.

Make no mistake: *the purpose of these events is to generate demand for your software and close business.* As a software vendor, the real value happens in two places: first, the main stage keynotes where potential buyers hear strategy and see announcements that get them fired up to do business with you; and second, in the side rooms where executives and others from both the vendor and buyer meet to talk about how to get a deal done. Each format communicates brand and solution-level values to your audience, whose curiosity is piqued as they enter breakout sessions. By the time breakout sessions begin, these participants are already prequalified for interest in sessions like "A Security Guide for Rockstars: Keeping the Bad Folks Out and More," and "Four Myths and Four Truths About Customer Journey," and so forth, which are mostly designed to get customers excited to buy or retain specific products by focusing on customer education.

Note

This is not to say that presenters have ulterior motives; most genuinely want to edu-
cate and inform so that customers will return from the conference with value—but
remember, as we mentioned earlier in the chapter, everyone is in sales.

Sometimes, breakouts like these are aimed at existing users; other times, they are focused on specific user roles, industries, or simply top-of-mind topics in the industry.

It is worth noting that these conferences, *when done right*, can offer massive ROI to your company, even with large chunks of the conference budget going toward hosting technical breakout sessions and putting on concerts (which, of course, have VIP areas for prospects!). Although it is easy to be distracted by the bright lights, star speakers, and the fancy venues, these conferences can be semi-nal events for enterprise software vendors and their communities of customers and users. You could call them the "Black Fridays" of enterprise software, which is doubly useful as a metaphor because it emphasizes how important good execu-tion is to make these events useful investments.

For that reason, it is worth a product manager's time to participate. How you participate will depend on the type and size of conference. For smaller events, you might be asked to take an outsized role, such as delivering a keynote on a given topic related to your area of expertise. For larger events, you might be invi-ted to host breakout sessions or participate on panels, lead a technical training, or demonstrate a product feature on stage. (We won't cover how to lead a good breakout session in this book, except to say that there is no enterprise software product manager who couldn't benefit from a public speaking course.[2]) All of that is well and good but falls into the category of "things you are doing at the conference while you are not helping drive your business."

These conferences bring together large numbers of customers and prospects, and as a product manager, you should take advantage of this. You will never have a better opportunity to interface with so many customers and prospects in one place. If your sales team is on top of things, it will reach out to you to have you meet with its biggest and most impressionable accounts. You can bounce from meeting to meeting for days, helping to answer prospects' questions and securing business for your field teams. Going into a large conference, make sure that you and your team are all clear on how you want to talk about your product strategy, your roadmap, and any hot-button topics that your customer base might raise with you. It isn't uncommon for a single customer or prospect to meet with two different product managers working on the same products at two different times, so you will want to coordinate and make sure that you are consistent in your approach. This is a huge opportunity to help your product's year go well in terms of bookings and retention, so make it count.

Your sales reps at these conferences will be frazzled, running around from account to account, trying to make sure that everyone is having a good time and is where they are supposed to be to hear the message that will help them see the light and want to do business with you. We call it "summer camp for sales" because they are jetting about seeing this contact or that contact, much like the excitement of a young kid returning to summer camp after a school year to see his or her old camp friends, all in one place again after having been apart for nine months. On the other hand, if a conference is make or break for your annual bookings targets, we can call these conferences life or death for some of your sales reps, who will need customers to have a positive experience and meet the right people (that's you) to help them decide to license your software.

The most important thing to keep in mind is that any conference is about winning new business and/or expanding within the customer base you already have. When you begin to look at these events that way, you will find ways to engage with your sales team so that you can best take advantage of these opportunities. Educating your customers is a good and worthwhile part of your time at conferences, but that won't pay for these events.

---

2 For more guidance on this topic, we strongly recommend this book by engineering management leader Lara Hogan: *Demystifying Public Speaking* (http://bit.ly/2C9OttH).

DEALING WITH BAD BEHAVIOR

One of the challenges that a direct sales model presents is compensation structure. As a product manager, you likely won't have much control over how your sales reps are paid, but it is critical that you at least understand the mechanics of it. Your leadership team will have designed what they believe is a good system of incentives and restrictions to ensure good behavior on the part of sales (e.g., not allowing contracts to be front-loaded to produce higher commissions). On the whole, however, human beings are pretty good at finding loopholes in any system designed to restrict them, and sales reps can make an art form out of this subtlety. Reps find the loophole, management closes the loophole, and then reps find the new loophole that was created when the old loophole was closed. On and on, *ad infinitum*.

There's nothing wrong with sales being "coin operated." That's literally their job. In fact, there are some situations in which this can work to your advantage. If your product is aimed at a sales persona, for example, sellers' willingness to "hack" their way around standard operating procedures in order to *do what works* can be an opportunity. In the early days of Salesforce, for example, one critical adoption path was salespeople effectively using Salesforce as a shadow CRM solution to their companies' existing incumbent vendor, which they just found painful and slow (if they had any at all!). Eventually, Salesforce was hosting CRM for lots of companies that had not yet signed official contracts, but who eventually would. In a similar way, thousands of companies are dealing with this today regarding Bring Your Own Device (BYOD) policies and employees' use of personal email addresses to handle company files. The bottom line is that if your product is truly, demonstrably better, salespeople are a great entry point.

The reason we mention this here is that you might find yourself in a situation in which you catch a sales rep doing something that seems odd, like structuring a contract in an unconventional way. Even though you and the sales rep both want the customer, and even though you are both judged using similar metrics, the fact that you are not paid in the same way is critically important. Most sales reps don't think beyond the current quarter, which means that you might need to be the one to do that. *As a product manager, you must be concerned with both the current quarter and the long-term health of your business.* A front-loaded three-year contract might help a sales rep get paid more commission up front, but it will lead to lower revenue for your product in years two and three, and make doing a larger deal at renewal time more difficult when the customer points out how inexpensive the product was in years two and three of the last contract. If this needs to become a conversation between management on the product side and on the sales side, that's okay, but do not let sales mortgage your product's future just to get paid today.

Hopefully we haven't made it sound like working with sales is a chore. It can be tremendously rewarding to participate so close to where your company makes its money, to see a prospect go from skeptical to convinced, or to help secure a large renewal from a happy customer. There are some caveats, but you will find the right balance between yes and no as you learn the structure and culture of your sales organization.

Perhaps the best benefit of working with sales is learning how large enterprise deals are done. As you progress in your career in product management, this experience will prepare you to lead your business at the highest levels of your organization, making the time you spend helping sales potentially very valuable indeed.

COMMUNICATING WITH SALES

The language of sales is completely different than the language of development or design; quite literally, you might often feel like you are speaking a foreign language when sales begins talking about linearity, upside, cross-sell, and marketing qualified leads. When it comes down to it, your average sales rep cares about acquiring new customers, growing spend with their existing accounts, and hitting their quota. Although they approach the world different than you and your engineers do, learning how to communicate effectively with sales will help unlock one of your most valuable and rewarding partnerships and prepare you for additional leadership opportunities in your career.

Most reps don't know how to translate "what a feature does" into "how a feature wins" (i.e., the value proposition), so help them get there. In short, you are going to need to dial back the development-speak and focus on how you expect customers to actually accrue value from your product. Remember the old marketing mantra that "people buy benefits, not features"? That might be marginally less true in enterprise software than in consumer software, but it still generally holds. When talking to sales about your product, or about your roadmap, make sure to tie everything back to how your customers are getting value from their relationship with your product. Here are some ways to do that:

Communicate using case studies and references

On that same note, a particularly effective way of communicating with sales is to speak in terms of case studies and customer references (both of which are essential to winning in enterprise software). What we mean here is to answer questions from sales by providing examples of customers who have been successful using your product. For example, if sales passes along a question from a customer who demands to know about your security protocols, answering the question directly is good. Even better, though, is answering the question *and* sharing the names of a few customers who are particularly demanding and have been satisfied with your solution's security offering. If sales asks how customers do marketing attribution using your product, send them a few case studies that show how actual customers demonstrated ROI using your product to decide how marketing budgets are spent.

Recognize and have empathy with the fact that what is happening in your product next year does not matter, more or less

This sounds a bit like it flies in the face of our earlier point about prospects buying your roadmap, so let us explain. That remains true; to the prospect, the future matters. Our point here is that *as far as sellers are concerned*, if they cannot sell it at least as roadmap this quarter, or this year, it may as well be happening next decade. Don't count on vague promises impressing your sales team. If you are not planning in the short-term to deliver a feature that they claim to need to win a big deal that is at the wire, hearing that it "might" happen in the next year or two will help but little.

Be clear about how your work will enable them to win against specific competitors

At the end of this chapter, we have included an example of a "battlecard," which is a document that product marketing and product management create to summarize the tactics that sales can use to deposition a pesky competitor. Whether you have handy battlecards or not, your communication with sales should be focused on helping them win, which often means explaining exactly how a brand using your software can do things that a brand using the competitor's software can't do. Sometimes, you will be asked to step in and help a customer or prospect understand how your product is unique in your market; other times, sales will come to you to help them assess a competitive situation. In either case, be explicit not just about differences in product capabilities and specifications, but also in the unique benefits and value you provide. You have the benefit of getting to work with your *entire* customer base, so use examples from real-life interactions in which you saw a brand do something that they can do *only* with your product.

The language of sales is different; learn how to speak it

Ultimately, the language of sales is the language of business growth. We gave some examples of this language in this section. Learning what sales means when it talks about linearity or downsell helps you work with your counterparts there and to show them that you understand their world, but it also gives you the leadership tools that you need to put your work into context for executives, who are likely all accountable for sales metrics to one degree or another. It is one thing to say that "customers will love Feature X," but it is another to explain that "we need Feature X because it will help combat attrition in our mid-market accounts where we are struggling to deliver cross-sell with any degree of predictability."

Help sales get some insight into the reasons you aren't building their requests

This might not be possible every time, but transparency about your challenges and goals go a long way toward building trust with your sales team. Suppose that a sales rep is demanding a certain feature to help close a deal. When you can, consider letting him into your world and help him see the trade-offs that he is asking you to make. He probably doesn't realize that development and design resources are extremely limited. He probably doesn't realize that delivering on his request would mean deprioritizing something that your entire customer base is demanding. A little transparency and allowing your sales reps to step into your shoes has helped many a sales rep to see that the product strategy and roadmap are doing exactly what they should be, even if it means a little bit of short-term pain for your sales reps. Give them the full context of your decision and it will be the foundation for trust between you and your sales team.

Listen to the sales reps, not just sales management

It is not uncommon for product management to spend a lot of time listening to (and training) sales management, perhaps as part of an internal product steering committee. This is good, given that sales managers are close to the sales reps and their customers, but they are still a step removed, so it is vital to spend time learning from actual sales reps what they are hearing from their interactions with customers. As one former product manager put it, "This is something I wish I would have done. Managers are rarely involved in the initial value proposition pitch. They almost always come in during the later phases of the sale; they're closers. Front-line salespeople are the ones that are in the trenches hearing the objections."

Summary

Organizational knowledge is a sprawling, highly company-specific corpus of knowledge that is critical for any successful enterprise product manager to master. This is even more the case in larger companies that contain huge amounts of institutional knowledge. We recommend focusing first and foremost on development, design, executive, marketing, and sales teams; understanding how they work, what their incentives are, and how product management most effectively interfaces with them. This is not an exhaustive list of our important counterparts, of course, but these are the most common teams that Product works with in many enterprise software companies.

PM Profile

Name: Jim Lynch

Title: Senior Product Manager, Salesforce Commerce Cloud

How did you get started in product management?

Before product management, I worked in services, first as a customer service rep, then implementation consultant, and finally a solution engineer. In all these roles, the root of the job is solving customer problems on a one-to-one basis. I started to look into product management as a way to start helping customers on a broader scale and as an opportunity to solve their problems and help them be more successful. I also took a year off for business school and wanted to find a role that would help me apply what I learned there within the software industry that I love being part of.

Are there particular backgrounds you think are especially valuable for enterprise product managers?

I've seen successful product managers come from almost every other area within software companies. What people learn in engineering, services, marketing, etc. can all be valuable experience for growing into a product management role, and I think teams comprised of PMs from each of these different backgrounds tend to complement each other well. In my opinion, the most common trait that is important for PMs is being a strong communicator to a varying group of different stakeholders. That is, they're able to describe product needs at the level of detail and specificity that an engineer needs to build it correctly, while also being able to communicate to customers and prospects about the business value of some highly technical product where the benefits to them might not be inherently obvious.

What have been the most difficult parts of your job to learn?

Saying "No." Coming from a services background, my inclination was (and remains) to always be thinking customer first, and always wanting to keep them happy. In a product role, you simply can't say "yes" to every customer feature request that comes in—your team almost surely doesn't have the capacity and the product that would result would become an incoherent mess. Learning how to listen to the request, dig into what the root use case or goal behind the request is, and try to identify the themes of all the requests you're hearing has helped make this a little easier. Customers are never excited to hear "no," but if you can explain your rationale and be open with them about the process, most are reasonable about it.

When have you felt the most fulfilled as a product manager?

The most fulfilling part of being a product manager is anytime I'm working on a feature or product that customers get excited about and can't wait to use. Being able to collaborate with engineering to build things that have an impact on our customers' success, and makes them happy to be working with my company and using my product, is awesome validation of all the work we do as PMs and product teams.

5

Product Knowledge

Product knowledge encompasses much more than simply knowing how your product works inside and out; it also enables you to build useful requirements for engineering, create a product roadmap, and understand when it is time to end-of-life a product or feature.

We expect that when many new product managers first begin exploring this book, they are going to flip to this chapter first. And we understand why. "Product managers" and "product knowledge" intuitively seem to go together. Indeed, they do, but this realm encompasses so much more than just knowing a lot about how to use the product one manages.

Product knowledge goes far beyond simply understanding what's in the product. It also encompasses the processes that exist *around* the product, as well: the planning and development, the users and customers, the lifecycle and roadmap. In this chapter, we're going to talk about all of these things. But the first thing we want to address is empathy. Cultivating empathy for the users and customers of your product is the paramount task for a good enterprise product manager newly on the job. And one of the best ways of doing that is by putting yourself in their shoes.

Learning Your Product

It's pretty obvious that any product manager must know her product inside and out. That's a tall order for almost anyone, especially if you're stepping into a product manager role for an already mature, far-reaching product. After all, unless you worked at or were a customer beforehand (and those moves do happen occasionally), it's unlikely that you would've had firsthand experience with, for instance, a particular vendor's enterprise Software as a Service (SaaS) platform before.

Fortunately, there's nowhere easier to learn the ins and outs of a product than on a team composed of the folks who built it. In our experience, the best way to learn your product is to use it, as closely as possible, in the same way your customers do. Your sales or QA team will have a demonstration account or environment that they use when presenting to prospects. If you are lucky, you might even have an entire team responsible for creating and managing demonstration accounts and dummy data. If your own company uses your product, even better. Get access to an environment. Deploy the product yourself, if possible. You want as much of the experience that your customer sees as possible.

In a team that we observed, the product managers and user experience (UX) designers were struggling to empathize with end users, which made design cycles longer and much more painful, ultimately affecting the time it took to bring new features to market. The two teams decided to meet every Friday afternoon, over potato chips and bourbon, simply to *use their product*. They invented scenarios based on interviews they had led with customers, and then tried to address those scenarios, just as an end user might. As members of the combined group had questions, someone in the room usually had an answer. By sharing, everyone learned together, and within a matter of a few weeks, product knowledge among all participants had multiplied significantly. As participants led conversations with customers, those conversations became more valuable and relevant, cutting down design time and giving the product managers involved better access to use cases that they could share with sales and marketing. The product and the company ended up benefitting from the growth of empathy and trust that comes from product knowledge.

Does this mean you need to be a Jedi Master with every module, report, and feature of your product? No, it does not. A good level of product fluency to shoot for in the beginning, in our opinion, is becoming "handy." A "handy" user knows how to perform basic tasks and can demonstrate the main functions in a customer setting. You have read the documentation (yes, really) and know your way around. On a technical level, you should also have a pretty good idea of how your product works under the covers. You do not need to know the codebase, but understanding the basic system architecture is a good idea.

The Product Lifecycle

Products, like anything else, have a lifecycle. Sometimes, you can predict them reasonably well, but usually, you can't. From launching brand new products and features, to transitioning into a sustaining, maintenance phase, to the eventual product end-of-life, understanding and planning for an enterprise product's full lifecycle is critical to give confidence to your customers and the rest of your team that you have a plan for what comes "next."

Product launches get all the attention. Across the tech industry, product launches are seen as opportunities for "growth hacking," or attracting as many eyeball-seconds of attention as possible to encourage adoption. Tactics range from the famous invite-only list that helped Instagram hoist itself into prominence, or Drew Houston's MVP demonstration video for DropBox (*http://tcrn.ch/2EgA4cy*) that propelled its beta waitlist into the tens of thousands. There are thousands of resources from folks much smarter than us about how to growth-hack your product into the next Facebook, and that's not really in our wheelhouse. (Just a warning: some of those tactics are much more obnoxious than others.)

For most enterprise products, launches work quite a bit differently. As the saying goes, one does not simply "growth hack" their way into the enterprise tier. Meaningful penetration into and adoption by most enterprise customers requires a sales process that's very well understood already. As we discussed earlier in Chapter 4, a lot of enterprise sales cycles are frequently 6 to 12 months long, high-touch affairs involving entire teams of people. In that world, if the first your customers are hearing about a new product is only when it becomes generally available, you've already lost valuable months of sales runway. That's why a standard enterprise strategy is to socialize new product marketing, information, and even demonstrations with existing customers and even prospects before the product is actually ready for release. (Early adopter or beta programs work in similar ways and can even help your team vet and iterate on the product before it even becomes public.) Assuming that you have established a lot of trust with your customers already, this can be a very effective way to "launch" the product with sales agreements already in place.

There are some exceptions, of course, but a lot of enterprise software is much more complicated than asking customers to go to a website and signing up for a beta. It's obviously just not feasible to offer a "freemium" model for complicated software solutions that require services and implementation cycles to get up and running. In an enterprise context, generating enthusiasm in your market and customer base about upcoming products comes long before the launch. Later in this chapter, we look at how product management can help our marketing colleagues do that with a clear roadmap planning process.

Most enterprise product managers are much more frequently launching new releases or features instead of completely new products. These differ from product releases in a couple of ways. First, net-new products tend to be revenue-generating (i.e., they are enabled for a customer as a result of a sales cycle), whereas new features or releases are more commonly revenue-neutral enhancements to existing product lines (though feature-level "upsells" are also common).

New feature-level launches are typically the result of long cycles of discovery research with customers. As with new products, it's rare that our customers are surprised by utterly new features. Account teams typically stay in regular communication with the customer, and product management usually shares the high-level roadmap plans upon request. If there are customers for whom a specific feature or tool are especially relevant, an obvious best practice is to work with the customer to gauge suitability for their use case. Sometimes, a customer will already be using another (heretofore noncompetitive) vendor to fulfill that need; in these cases, at a minimum, it's wise to telegraph that the feature is coming soon, and if possible, sit down with the customer to discuss it in more detail.

Release launches tend to result in more responsibility for product management than other kinds. Some organizations ship releases in cycles as short as two to six weeks. Others, especially big, mature platforms, might only do quarterly or biannual releases. Clearly, the latter case involves a great deal more planning around release marketing to make sure that customers, prospects, and the market at large are as aware as possible of all the cool stuff you've built. Because this typically involves mass customer and market communications, this again falls mostly into the domain of marketing, but product management has a critical role to play in ensuring that information about the release is clear, understandable, and relevant to customer needs. Marketing collateral about releases should, like other product-related information, be vetted with product management and, ideally, created with some level of input. Short explainer videos adjoining your release are an excellent way to quickly and cheaply put out consumable information for a user base that is unlikely to wade through a full release notes document.

In many cases, launches themselves are really more about marketing than anything else. They are the culmination of long planning and hard work that hopefully translate into business results. You might want to skip down to the section "Measuring Success" on page 104 at the end of this chapter, in which we cover some example key metrics for product managers to track when a product or feature "ships."

Writing Stories and Requirements

A large chunk of a product manager's job boils down to good writing. In fact, we'd say that the ability to express oneself well in clear, concise, and mechanically sound prose is among the top five criteria for hiring a new enterprise PM.

Writing clearly understandable user stories and requirements is a critical skill for any product manager to develop. There are few widely accepted frameworks for doing this, and, in fact, many large, mature organizations like our own have in-house templates that are encouraged (and enforced to varying degrees). In most cases, your best bet is to go with the flow and learn the prevailing system before introducing any radical changes. Nevertheless, most good user stories will include some variation on the following:

Title: [What this story is about]

    As a: [Role or function of the user in question; example: business analyst, database administrator]

    I want to: [The immediate goal from the user's perspective that this story seeks to fulfill]

    So that I can: [The reason why this user wants to do this, in case it's not obvious—what business value this achieves]

An example might look like this:

Title: Share access permissions at the "user type" level

    As a: Administrative user

    I want to: Grant access to sales pipeline forecasts to users based on their type, not individually

    So that I can: Granting access to users individually is time-consuming and tedious. By selecting all users of a certain type, I can save time, and also enjoy more peace of mind that all users are properly provisioned access, because one-off exceptions will be rarer.

A good, admittedly basic user story like this often beats a 15-page Product Requirements Document (PRD), because it focuses on what the user is actually trying to achieve rather than how it's accomplished, per se. It keeps the story bounded within the context of a user's actual task, rather than what the product manager wants that person to do.

There is a wide array of supplementary information that can be included on a given user story or requirement. In some cases, multiple pages of a PRD might actually be needed, though not in others. For example, when you're designing the beginnings of what will be a "foundational" product on a platform, which customers will rely on as a jumping-off point before possibly using other add-on modules, a PRD might be useful as an internal reference. Sometimes, wireframing a prototype is an important input that can help developers and designers visualize what you're trying to describe.

More important than sticking to any specific story or requirement template or structure, however, is taking the time and attention necessary to write clearly. You should assume that your audience has a minimum of outside contextual understanding of the target user or his task. That audience—members of your development team, designers, and other product managers—surely *do* possess that contextual understanding, but it might be a shade different than yours, which is a good thing! It might help them contribute insight into the user's goal that you hadn't considered.

Building the Product Roadmap

"The roadmap" is often one of product management's most visible pieces of work, and it certainly gets the most airtime. Like the office pizza order, everyone wants to know what's on it and has opinions about what should be in it. As the product manager, everyone will expect you to know and explain. So let's start at the beginning: what's the purpose of the roadmap in the first place? What function does it serve?

Like Janus, the roadmap has two faces: one internal, and one external. They each serve important, but markedly different, roles, as illustrated in Table 5-1.

Table 5-1. Product Roadmap Functions

Internal	External
Clarification on high-level goals	Communicates new upcoming capabilities
Aligns product management and development	Demonstrates customer responsiveness
Informs resource allocation	Advertises vendor engineering quality
Clarifies company business model	Branding and marketing thought leadership

Internal	External
Guides longer-term planning process	Solution/industry community-building around product

The roadmap that a company shares with its customers and telegraphs to its market is both a marketing tool and, in a very real way, a part of your product itself. Not only does the high-level roadmap a company promotes externally entice future customers by telegraphing sexy new capabilities, but it also instills a sense of *owned value* on the part of current customers. In our world of enterprise SaaS, customers who subscribe to our platforms are, effectively, not only buying our product offering as it exists today, but are also *investing in our product development process itself.* After all, the nature of multitenant SaaS extends to all customers the most up-to-date features and capabilities of the product, which intuitively means that the value it delivers should compound over time. Therefore, the roadmap we share with customers must demonstrate some return on that investment in a form they appreciate as meaningful and clear, but that new customers can still recognize as attractive. Think of it as both customer retention *and* acquisition.

On the flip side, your internal product roadmap is an invaluable organizational aid. It illuminates a path to achieving the organization's business goals (which should come first). It gives clarity to everyone in the organization (or at least in the relevant groups) about your immediate destination and overarching goal, which itself is a critical ingredient of leadership. For our purposes, it aligns product management's vision for the product with a development plan for getting there. If product and development start off on the same page about where they're headed, and in what sequence, interim deliberations can center around the more productive realms of "how and where" instead of interminables like "why and when."

This is why it can help to use the following pattern in your external and internal roadmaps.

Externally, you want to underpromise and overdeliver. Give yourself an extra sprint or two, or perhaps a quarter, of leeway for big features or enhancements you plan to release. This way, if progress slows or unexpected issues arise (which they invariably do), you have a cushion, and if they don't, you can release early and look great.

Internally, it's useful to set baseline and stretch goals for your roadmap, which will probably have a different format than its externally facing counterpart.

In this model, product management and development agree to a certain set of broad goals for the planning period in question (i.e., the quarter). These might be represented by a series of epics, which are then broken down into stories and allotted to the available sprints. In the event that development has some extra cycles at the end of a given sprint, it's useful to have a handful of one-off or miscellaneous stories in your backlog that are preprioritized and ready to peel off. Sometimes, you might want to just let development choose what it would like to work on beyond the baseline. Many developers are inherently curious and innovative, and examples are legion of useful new features or enhancements growing organically from developers' side projects. (Unfortunately, there are also lots of examples of the opposite too, so it's a mixed bag.)

ALIGNING ROADMAPS BETWEEN TEAMS

It is frequently the case in our world that enterprise products (and thus enterprise product managers) are organized into many separate teams within a single company. If these products are wholly separate from one another (e.g., Atlassian's JIRA versus Trello versus Bitbucket), with a wide dispersal of user types and buyers, there might be little or no reason to tightly integrate their roadmap processes, and they might even rarely interact. If, on the other hand, an enterprise SaaS platform supports multiple different modules or product offerings (e.g., the Salesforce CRM platform or Adobe's Marketing or Creative Clouds), these teams will inevitably be faced with the challenge of aligning solution-level product roadmaps.

If you watch HBO's *Game of Thrones* and saw the "Battle of the Bastards" (*http://bit.ly/2G33nQ7*) episode, you probably have a pretty good idea (*http://bit.ly/2EhxarW*) of what those meetings look like. They usually start off like your typical single-product roadmap meeting—sort of a collection of wish lists from different teams—but then also mix in organizational rivalries and politics (and the dreaded HIPPO[1]), and it quickly becomes a mess. It's easy to see why this is so: if product teams A, B, and C are all asked to collaborate to hammer out a solution-level roadmap, but each is evaluated on team-specific product performance and revenue targets, that sort of hurts the incentive to give much ground when allocating development resources, doesn't it?

It helps to first define what we mean by "aligning roadmaps." One level of alignment might mean simply agreeing to common themes for product releases,

---

1 "Highest Paid Person's Opinion"

which is relatively easy. Another, slightly more complicated level might be a cross-product project or two, like some form of two-way integration. This is more complicated because it requires some identifiable user value for both product teams, but because integrations tend to be encapsulated, finite projects, it's relatively straightforward to scope such a project and dedicate resources to squaring it away.

When each product team supports discrete, salable offerings, aligning projects and resources can begin with fundamental business goals. What products sell? For which are we projecting future growth, or which are we betting on? Aligning development resources, prominence, and roadmap primacy based on these factors is pretty straightforward for the leaders of product organizations. When clear direction from leadership is absent, however, it's not uncommon to see sheer inertia beget big imbalances between teams' revenue contributions and dedicated resources.

As just one example, we saw at one large enterprise vendor how solution-level leadership failed to take action to align product-level resources after the big acquisition of a complementary competitor. Although both acquirer and acquired product teams discussed ideas for integration and even complementary new features, the performance of each team remained measured by standalone product sales. This meant not only that there was little incentive to invest heavily in deep integration, but that the team that did so could actually be penalized because the sales strategy for each remained unchanged. Product leaders tore their hair out pleading with executive leadership, describing the customer and user problems that further integration could solve, but the executives, who in turn were hyper-focused on quarterly sales results and lacked deep experience in the target industry, were unmoved. The situation did not resolve itself. More than a year later, the acquired company's development team still had a higher headcount and more development resources than the acquirer's team did, despite the latter posting almost double the revenue. It was later revealed that this was partly due to the company's finance department, which struggled with incorporating the concept of Annual Contract Value (ACV) for the latter's SaaS model into its planning metrics. In comparing the two products, finance was basically comparing perpetual license revenue for the (acquired) on-premises software versus the latter's recurring revenue. As such, finance suboptimally budgeted resources for each, and most of the executives just didn't know enough to argue the point. (This makes your authors cringe just to remember.) In the end, both products eventually withered and died on the vine.

This is both a cautionary tale for enterprise product managers as well as (we hope) an encouraging one for smaller, more nimble enterprise competitors. Large enterprise software companies are not always known for their responsiveness or agility, and bureaucratic hindrances like these are legendary in the industry. It seems hard to believe, but even today, many legacy software incumbents still struggle with the very concept of SaaS. Make sure you avoid the same mistake.

THE ROLE OF CUSTOMER INPUT ON THE ROADMAP

Considering how much emphasis we often place on gathering feedback from customers, it's no surprise that their input has a strong influence on our roadmap. But that influence is not definitive. Customer knowledge is often more a matter of building empathy for users and customers as a means of *inspiring* better products than simply asking what they'd like to see. Although it's important that your roadmap reflect certain customer-requested feature enhancements and tweaks, it cannot become the customer wish list.

There are several useful ways to collect customer roadmap-specific feedback. One of the most popular in the enterprise world is the Customer Advisory Board (CAB), which is a pretty standard feature for most enterprise companies that we discuss further in Chapter 6. The CAB brings together representatives, as senior as can be recruited, from your company's customers on a regular basis for broad-ranging discussions on many topics, including your roadmap. Although typically not a sponsor of the CAB, product management is often called into these meetings for high-level roadmap presentations and to solicit feedback. Because a CAB typically represents a collection of executive customers, this is a golden opportunity to test and validate business value propositions for new features, modules, or messaging.

Here's another tactic for roadmap-related feedback: in your next meeting with customers (whether users or executives), try bringing a few stacks of *Monopoly* money. Then, distribute some budget to each person in the meeting (yourself excluded, of course)—say, $200. Put down on paper a range of different product themes, features, or topics that represent areas for improvement or even specific enhancements you're considering for the roadmap; 10 is a good number, obviously depending on your product. Each product feature/theme gets its own sheet of paper arrayed on the table. Then, ask each person to allocate their *Monopoly* money "budget" according to their desired preferences. They can put all $200 on just one topic or spread it out (un)evenly across many of them, or even bargain with colleagues to "buy" one feature or another. (There are any number of variations you can try on this game!) Track the results, particularly by user role (developer, business user, analyst, executive, etc.). You might be surprised by the results. At the very least, fun, engaging exercises like this can provide fertile ground for jumping off into valuable conversations about specific product feedback.

Always remember that it is not your customers' job to make you money, or to build your company's empire. They are looking out for their own challenges, both short and long term. This might be an obvious point, but it's worth a reminder whenever you consider customer feedback on your roadmap.

Politics and the Roadmap

Unfortunately, the nature of the product roadmap also makes it inherently political at almost any company. Obviously, you want to avoid that to the extent possible. Here's what you can do when it is becoming politically charged.

There are two primary vectors by which politics infests the roadmap: either the executive mandate (executives want what they want), or by trying to please too many stakeholders at once (the dreaded "roadmap by committee"). Both can be a big, heavy anchor on your progress.

First, make sure that you're not the anchor. What if you're wrong? Could your strong-willed executive or recalcitrant committee member actually be right? Consider the issue from their perspective while ignoring as much of the personalities involved as you can. We know, it's difficult. Put your ego (or machismo) aside and try it anyway. The ability to analyze a situation objectively is one of the most critical skills you will develop as a product manager (if you haven't already).

Positive you're right? Okay then. Next, there are few counterarguments as powerful as a good old-fashioned business case. If you're arguing over a feature inclusion on the roadmap, for example, game it out: what will Feature X get you? What will it cost, either in terms of development cycles or additional service costs? Is it a revenue driver or just a whiz-bang demo gimmick? Get input from well-known sellers if you can and present your case in hard figures. Keep your language cold and concise; let the cost/benefit breakdown do your persuasion.

Alternatively, you could try a classic flanking maneuver and coopt the requested (/demanded) roadmap inclusion into a thematic package.

Thematic cycles are an excellent release practice in general, as we've discussed earlier, and can present an opportunity for you to sneak the demanded roadmap item into a longer list of more desirable work by clumping them together into one broader topic. This might even present an opportunity to minimize the item's prominence or development level of effort later on down the road.

Note that you might be able to turn roadmap disagreements like this into an opportunity to reexamine the roadmap planning process itself. The executive mandate and roadmap-by-committee approaches are classic antipatterns to sound roadmap planning. After the present question is past, bringing up your roadmap planning process itself for reconsideration can be a great opportunity for product leadership.

Lastly, it's worth recognizing that some level of politics is probably inevitable. No large organization of humans operates without it. Over time, people either learn to swim with the tide at a company, or they don't. Software development is a team sport. Sometimes, the best response is to swallow your pride, "disagree and commit," and work on your breaststroke.

Planning Process

Note that the roadmap is not a Soviet-style five year plan, nor a forced march. When you plan out a roadmap too far in advance or are too inflexible about sticking to schedule, you degrade its value and limit your organizational agility. Frankly, it's probably impossible to predict much of anything beyond 12 or 18 months in the tech industry (if even that far). The variability of your results naturally increases as you extrapolate into the future. For a customer-driven organization that relies on your roadmap for its business plan, a one-year product roadmap is about as far as it usually makes sense to plan. (Keep in mind that your product *vision* and *mission*, of course, are quite a different story.)

Most companies do some sort of annual planning process. The same kind of cadence is appropriate for planning your product roadmap. The process by which any company plans out its roadmap inevitably involves a bit of sausage-making, but what follows is our suggested outline for delivering an annual roadmap that is customer driven, advances company goals, and satisfies all of the main internal stakeholders:

One quarter before the next annual roadmap is due:

1. Product management begins by reviewing the developing product vision. Does it still hold up? Are you executing on the company mission? If not, start by clarifying, as soon as possible. Assuming so, proceed.

2. Data review. Get your hands on sales data from the last several quarters. Review total user growth and, if available, user engagement data (analytics). Sift through these to understand if there are specific product challenges the data is telling you to solve. (To go into more detail on this, skip to the section "Measuring Success" on page 104 at the end of this chapter.)

3. Get an idea of the company's sales/revenue goals for the upcoming year. This is one of your starting points. What does the company need from the product roadmap to achieve its overall goals?

4. Here's the fun part: start thinking big. If you were dictator for life, what would your big stretch goals be for the product this year? Are they new revenue drivers, or basic enhancements? Do they fill competitive gaps, customer requests, or are they greenfield ideas? Importantly, do they help the company meet those business goals from the previous step? Come up with a list of big goals as well as a longer list of smaller enhancements. (This is a good opportunity to mine all those user stories you've written and thrown into the backlog all year.) Put this all into your rank-ordered product management wish list.

5. Schedule a planning meeting with development. This can either be with the entire development team or just the development manager(s) plus whoever wants to come. Send over the product management wish list ahead of time and ask that attendees review it in advance and come prepared with feedback. *Ask that development have its counterpart wish list, if one exists, prepared to review.*

Two months before the roadmap is due:

1. Planning meeting with development. Development's list should look familiar—a combination of product features as well as required maintenance, like resolving technical debt and platform upgrades.

2. Triage. Hash out and prioritize which items must be done soon, which can wait, and which best help achieve company goals. Balance and prioritize as necessary, and ensure that epics are reasonably sized to the cycles available. Finalize a quarter-by-quarter Candidate Roadmap at the epic level that is tentatively endorsed by both product management and development.

3. Raise the Candidate Roadmap to the Product Leadership Team (PLT) and schedule a meeting to review it. (Check out Chapter 2 again for more about the PLT model.)

Six weeks before the roadmap is due:

1. Review the Candidate Roadmap with the PLT. Presumably, product management is already familiar enough with the PLT's view that nothing in the Candidate should come as a complete surprise. Take PLT feedback on features included in the Candidate as well as their scheduling and prioritization.

2. Review PLT's feedback with development to ensure requested changes are within scope. If not, surface immediately. Reshuffle the Candidate as necessary.

3. With its feedback incorporated, the PLT provisionally accepts the Candidate Roadmap.

Four weeks before the roadmap is due:

1. "Socialize" the Candidate Roadmap internally with sales and marketing (and any other critical stakeholders who are not on the PLT) and gather feedback. Do these teams see any *glaring* omissions or have suggestions for tweaks? At this stage, major changes in epic inclusion or scheduling should be avoided unless deemed truly necessary.

2. Product management acts as a router for the inevitable back-and-forth of tweaks to the Candidate Roadmap between sales, marketing, development, the PLT, and product management itself.

3. Product management and marketing begin outlining roadmap presentation material, if this hasn't happened already.

Two weeks before the roadmap is due:

1. Final review and endorsement by the PLT. All members are accountable. If there still isn't consensus, resolve ASAP. Disagree if necessary, but commit.

2. Product management and marketing develop customer-facing and internal-facing versions of the Final Roadmap. Meetings are scheduled to present the Final Roadmap internally.

Due date:

1. The Final Roadmap is shared with relevant internal teams.

2. Product management is prepared to present on the Final Roadmap to sales teams, with customer-facing versions developed and blessed by marketing.

3. Product management and development create stories and requirements for executing on the first sprint/quarter of the Final Roadmap.

Congratulations! You have a roadmap!

Now, at this point, you're probably thinking one of the following things:

This is ridiculous! I've never seen a roadmap planning process this clean!

Or,

This is ridiculous! I've never seen a roadmap planning process this bureaucratized!

Both are, in truth, completely fair reactions. This is an idealized template—most roadmap planning in the real world involves a great deal more chaos than this lets on. In some organizations, "roadmap planning" often amounts to each product manager essentially showing up in a room and adding an item to a list.

By contrast, the advantage of sticking to a process like this one, particularly in a larger organization, is that it avoids a "peanut buttering" approach to roadmap planning (spreading around to fit everyone's wish list). By using the company's business goals as a starting point and vetting the approach to accomplishing them early with the PLT, you can better avoid the distractions of side issues and pet projects. It ensures that the product roadmap meets the organization's business needs while also staying customer-driven.

Measuring Success

Zooming out for a moment, let's consider a broader question: How do you determine how the product is doing? What constitutes success or failure? What's on your dashboard? Let's talk about how to measure product metrics to determine—and define—"success."

Suppose that your product hit 110% of its revenue target for the year (whether in terms of new bookings, new ACV, or whatever metric you use). Most people would consider this a successful year, and perhaps it is. Certainly, this is the number that many higher-ups will look at before moving on. But there's much more to the story here that's worth our attention. This has to do with the metrics of your success.

What gets measured gets managed, right? It's tricky to offer one-size-fits-all suggestions for measuring your product, because the varieties of enterprise software are so vast. That said, we would propose three relevant metric areas that will help tell a much more holistic and nuanced story about where your product is strong and where it could improve. These are **sales**, **users** (e.g., user growth), and **engagement** (e.g., actual user adoption and usage). Those latter areas, users and engagement, are typically much easier for a SaaS vendor to measure and track than for on-premises software.

On balance, it's usually good news to see growth in all three of these areas simultaneously. (And if you do, congratulations!) It's more than likely, though, that you see room for improvement with one or more of them. Given the industry and highly role-specific nature of enterprise software, however, what we see is that different combinations of these metrics put into context can provide lots of varied narratives around the health of a product. What do we mean? Let's take a look:

- Consider a product with *high* sales growth, but *low* scores for users and engagement. What's going on? Well, your product's value proposition is resonating among the executives who buy the solution, and your sellers are demonstrably effective in delivering the messaging describing that value. The problem is that value proposition isn't as clear to those executives' direct reports. The product isn't solving the problems those users have, and as such, it's unlikely to ultimately solve the larger business problems, either. The question for product management is whether the product *really can* solve those users' problems, and there's just a gap in onboarding or usability; or, more ominously, if there is a more fundamental disconnect for the client company between the problems its executive leaders are trying to solve versus those of the rest of the organization. Either way, this is a dangerous place for you, as a vendor, to be.

- Next, consider a product with *low* sales growth, but *high* scores for users and engagement. This, again, could point to an internal disconnect for your clients. It's more than likely that the problems your product solves for users are not necessarily shared among those executives who hold the purse strings. Evangelism and brand marketing can narrow this distance a little by building internal champions, but fundamentally, product management needs to find a way to connect the utility that users have discovered to business value that executives recognize.

- How about if sales growth is basically meeting expectations, but you're seeing *low* user growth paired with *high* engagement? This is a classic pattern for a niche solution: a few users find your product extremely useful, whereas most don't. With a closer look, you're quite likely to discover that many of those users share a certain role, responsibility, or focus area. This is a strong signal of traction that you can use to research those use cases more closely to better highlight them with customers and help inform your messaging.

- Finally, what if sales growth is basically meeting expectations, but you're seeing *high* user growth and *low* engagement? A lot of people see "low engagement" and immediately become nervous, and we understand, but slow down. Many enterprise products really don't require high levels of user engagement to be successful. Perhaps users only feel they need one specific thing from your product to solve the problem they face, and can then go on with their day. This, too, is a strong signal that can help you build on that use case to broaden the usefulness of the product.

Not all of these frames will "fit" for your product's context, of course, and you should take them mostly as examples. Yet by tracking and interrogating a small group of core metrics like these, product managers can begin to reveal a narrative behind their product's strengths and weaknesses that can be extremely useful in brainstorming paths forward.

Planning the Product End-of-Life

There could be many reasons why you end-of-life (or "sunset") a product. Hopefully, a newer, better counterpart is up and running, and you're ready to force customers to migrate. Perhaps your business has run into trouble, or the product is simply a peripheral one that you've made a decision to stop supporting. Whatever the case is, product end-of-lifes are a reality for many enterprise product managers and are important to plan for closely. A good product end-of-life process leaves customers feeling supported by a well-organized business partner who they are very likely to work with again. A poor end-of-life process will burn bridges and leave a customer hanging who wants nothing more than to advertise their bad experience to anyone listening. Consider a good end-of-life plan a sound investment in customer retention.

When you sunset a product because it's being phased out or replaced by a new version, planning should begin internally between product management and the sales team. What follows is one possible plan for the product end-of-life that is reasonably high-level so as to allow for company-contextual factors. This assumes, of course, that the reason for your product end-of-life is not because of a business event at the company (like, say, going out of it...).

Your first step is to pick a drop-dead end-of-life date and stick to it. There are likely legal or contractual requirements that will come into play around when exactly you *can* sunset a product, and for whom and when, so obviously be sure to get with your legal team. Be sure you understand what customers are legally entitled to in their contracts and for how long—the last thing you want is to be surprised by an angry customer wanting to know why you're winding down a product they're actually still entitled to use for months after your end date. But assuming you're in the clear legally, stick to your date and don't move it. No exceptions. After you set this date, immediately inform your sales teams. Ensure that no new deals are initiated for the product in question, and try to figure out a way to modify or terminate any already in the pipeline. There *may* be exceptions for which it makes sense to sell a customer a product that is on the verge of being officially sunset; for example, if the customer's needs are immediate and short-term (and, well...they still want to buy it)—but they are rare.

After you've finalized your end-of-life date and it's been blessed by management, it's time to get down into the nitty-gritty of developing a sound migration plan for customers. For example, if you have a newer version of the software or an alternative product entirely, migration probably involves more than just throwing a switch; it could involve moving or modifying a large amount of data, downstream applications, processes, and integrations. In complex scenarios and depending on the product, this might even involve professional services requirements. Many firms in this position choose to offer services at a discount, or even free, to keep customers. (On the other hand, some choose *not* to do so in order to put some subtle pressure on those customers to leave. "Firing" troublesome customers, like those with low subscription value but exceptionally high customer service demands, is sometimes a good idea.) For cases in which you do *not* sell an alternative product to position to customers, partnering is an excellent idea. Rather than leave your loyal customers hanging, it's a great idea to strike up a partnership with an erstwhile competitor that you respect and can work with to move customers easily from one product to the next. Through collaboration with the development and services teams, it should be product management that drives the primary customer migration plan and then presents it for approval.

After you have finalized a migration plan, it's time to evangelize it. Call a meeting with the relevant account teams and explain both the end-of-life decision and the migration path. This is where the upfront investment in creating a clear, well-documented plan will save you a lot of headaches. Most questions can be easily answered by a basic migration document and handy FAQ.

The next step is to go back to our friends in marketing. Marketing will again be a key partner in at least two ways. First, marketing can lend valuable assistance in developing an announcement plan for the end-of-life. In some cases, marketing can actually drive the announcement, particularly when it's used to position an enhanced or next-generation product. Secondly, and at a minimum, marketing will need to purge any existing collateral of references to (or pitches for!) the sunset product.

With these pieces in place, you're ready to begin breaking the news to customers. How this is done obviously depends entirely on your customer relations, but often, account teams, especially for big clients buying expensive suites of software, will call in product management to explain the end-of-life decision and walk clients through the migration plan. (If you're low on frequent flier miles, get ready.) These customer visits aren't always fun, but we've personally found that few customers are taken by surprise by announcements like this, and often react well when a clear migration path has already been well thought out and presented in good faith. Be straight about why you're sunsetting a product and how you intend to assist the client in wrapping up the project. In our experience, adults always appreciate forthrightness and an honest willingness to help. In sunsetting major enterprise products, we've had remarkably few negative reactions to this kind of explanation. (Sharing stories about the negative ones will require buying Ben or Blair a drink first.)

If you're positioning an alternative or next-generation product, it's a good idea to have a great pitch and demo ready to roll. If customers who are losing a tried-and-tested old favorite product can see, touch, and ask questions about a new alternative, you already have a warm opportunity to retain a loyal customer, which is almost always cheaper and better than finding a net-new one. In fact, it's standard practice to offer customers interested in moving to your next-generation product a sweetener—again, professional services is a favorite for this, but attractive discounting works, too.

Product end-of-lifes are a fact of the software business and nothing to shy away from. They're not as much fun as launches, but if you're in this for the long term, you'll go through them more than once. Just be ready.

Summary

Product knowledge is about much more than simply being familiar with how your product works. That's important, of course, but don't worry about being the world's greatest expert on Product X. Rather, after achieving a skilled level of familiarity, expand to understand all elements of how your product is planned, launched, maintained, and, sometimes, withdrawn from the market. Writing clear requirements and building and evangelizing the roadmap for your product are key components of product knowledge. Be sure you understand the dual uses of internal versus external-facing roadmaps because these have big consequences for what each should look like. As you execute on the roadmap, measure your outcomes on more than one dimension to make sure you're not missing critical narratives about your product's performance.

Table 5-2. An example product roadmap overview slide

	Q1 (v1.4)	Q2 (v1.5)	Q3 (v2.0)
Marketing Analytics	• Real-time reporting • Improved navigation	• Advanced metric builder • Improved sharing of data between teams	• Table builder and curation for analysts • Consumption experience for marketers • Updated permissioning
Advertising	• Advanced bid optimization		• Integration with AdWords and Bing
Machine Learning	• Anomaly detection • Statistical formulas/ functions • Statistical correlations • Gains/lift charts for scoring	• Anomaly detection alerting • Rules-based attribution modeling • Decision trees and other visualizations	• Contribution analysis
Developers/ Partners	• Firehose (beta)	• Report API enhancements • Standard authentication	• Report API support
Integration		• Analytics – Target data integration • Shared profile (Raven) • Additional card types	• Seamless persona switching
Mobile	• App-centric user interface • Points of interest reporting and targeting • Mobile OS and browser detection	• Push campaign management • In-app messaging • App download conversion tracking • Dynamic tag management integration	• Full geo-fencing analytics • Screen/Visual overlays • Enhanced messaging (SMS, rich media) • Mobile ad network integrations

| Video | • Heartbeat video reporting
• Integrated video and ad reporting | • Live reporting of concurrent viewers | • Support for linear TV content online |

Building Better Products with Data

Whether you're in enterprise or consumer software, analyzing data to understand how people actually buy, implement, use, and retain your product is your best chance to build the right thing quickly.

In Chapter 7, we examine how customers do not always do exactly what they say they do. You might ask customers questions about their usage of your product, or the challenges they face in their respective industries, and at times they are likely to give answers which are (either intentionally or unintentionally) uninformed, biased, or just plain inaccurate.

We've discussed how observation is a good way to ensure that you are focused on what people *actually do* rather than simply their impressions or assumptions about what they do. The trouble with observation is that it does not scale very well. Even in enterprise software, where your user base is likely substantially smaller than highly successful consumer products, you might have hundreds of thousands or millions of users that you would like to observe. Watching a small group of users perform daily tasks can provide you with powerful, visual evidence of the strengths and weaknesses of your product, but unless you are watching at least several dozen users try to accomplish similar results, it is difficult to draw widespread conclusions from observation alone.

The missing link to help you understand broader trends among your users is *data*, which allows you to spot trends in your user base (and customer base) at scale rather than individually, helping to ensure that you are making informed decisions about product design, the roadmap, and more. The best enterprise product managers are capable analysts who can ask good questions (and pose hypotheses) and then answer those questions (and attempt to invalidate those hypotheses) using data.

The use of data to help inform product design and the roadmap is, frankly, an area where consumer software has somewhat outpaced enterprise software to this point. There are a few reasons for this. Most notably, the sheer scale of consumer software user bases makes gleaning insights from *direct customer interviews* more difficult for consumer software product managers, in lieu of analyzing programmatic usage data; whereas enterprise product managers typically have anywhere from a few dozen to a few thousand customers to choose from, making qualitative customer interviews easier, and programmatic usage data somewhat less representative of the market. An enterprise product team can also use the direct-sales relationships with those customers to find opportunities to interact. However, the best enterprise software product teams are making great strides and much better decisions with data and analysis. This is another case in which we should not let the consumer software people have all the fun.

Fortunately, the past 15 years have seen a rapid expansion of organizations' ability to capture, distribute, and analyze data of various kinds. In this chapter, we briefly cover three broad types of data that will be essential in your own analysis as a product manager. If you have ever felt lost amid countless possible paths or design decisions as a product manager, this chapter is for you. If you have ever wondered whether what you are hearing from one customer is applicable more broadly to your customer base (or potential customer base), this chapter is for you. Our aim is not to teach you how to do data analysis or prescribe methods for analysis (cohort analysis versus latency analysis, etc.); other books and blogs will fill in those gaps.

Product Usage Data

Which features in your product produce lasting usage? How do your most habitual users differ from occasional users? Which features correlate with renewal or upsell? How should we think about packaging disparate components of our product together? What little-used features can be retired to recover development and testing resources? You can answer these questions and many others by examining data around product usage.

The key word with product usage data is *granularity*. It is nice to know how many people have installed or logged in to your product, but this is useful only for tracking the highest-level trends in adoption and churn. Getting more granular, specific data on which features are used most, which settings really make a difference, and how users break out into discrete groups or clusters along their journey with you is what makes usage data so powerful.

Admittedly, product usage data is easier to come by for some products than for others. The emergence of Software as a Service (SaaS) has opened up software to the same methods of data collection that websites have enjoyed for decades. Thus, for SaaS products, collecting data on product and feature usage can be as simple as inserting JavaScript tags or API event calls throughout your product. For traditional on-premises or installed software, collecting granular, feature-level product usage data can be more difficult or even impossible at first, due to connectivity, firewall, or security concerns. This book is not the place to go into various strategies for implementing user measurement within installed software; rather, we simply aim to recognize that an enterprise software product manager who is leading an installed product might need to become a bit more creative to obtain this data.

Net Promoter Score (NPS), which purports to give you a way to track customer satisfaction with your product over time, also falls into this category. Again, where possible try to be both general and granular with your NPS and voice-of-customer efforts so that you know not just whether users love your product, but which aspects of the product keep them coming back for more, and (no less important) which features drive them to rage.

At its most basic application, product usage data can also be a great way to find customers to partner with for discovery and feedback (as discussed in Chapter 6). Let's say you are considering iterating on a part of your product or that you want to understand in more detail the value users get from a feature. Product usage data can point you to the heaviest (or lightest) consumers of a part of your product. As we just mentioned, you can then take advantage of relationships that already exist between account teams and customers to reach out to the relevant users and spend time learning from them.

Sales and Finance Data

As a product manager, there are a number of reasons why it is crucial to be intimately familiar with the data typically owned by sales and finance. You are responsible for the business that surrounds your product. Ultimately, although it might not be how you choose to express your product goals at any given moment, the Key Performance Indicators (KPIs) that matter most in nearly every case for enterprise software are bookings, revenue, and retention/attrition. In one way or another, you are responsible for each of these metrics. The best way to ensure that you are on track to secure growth and to become involved appropriately when you are not on track, is by working with basic sales and finance data.

From sales (likely via your customer relationship management [CRM] tool), make sure you can access and filter customer data by products and entitlements owned, annual recurring revenue (ARR), location, account team members, and a variety of other fields that your sales operations team is likely already capturing (or could begin capturing). This data will help you do things like model out your customer base by revenue, track the progress of key sales deals and renewals as they happen, and aid in win/loss analysis, among other things. Product managers who have access to query their organization's CRM data can answer all sorts of questions and be much more involved in securing the future of their business. As an example, we are aware of an enterprise software product manager who gets a data dump from his company's Salesforce instance containing a snapshot of his product's entire book of business. From this, he can look up how much any given customer is spending on his product, other products owned by any given customer, and more. This data has been instrumental in figuring out what the most lucrative cross-sell opportunities are, which helps him plan roadmap work to address those opportunities for customers.

From your finance team, you will want a frequent readout of your bookings, revenue, and retention/attrition as each quarter progresses. Of course, it is crucial to understand after each quarter whether you hit your various targets, but having a regularly updated view into these metrics throughout the quarter allows you to choose where to focus your support of account teams, as well as which marketing levers to pull to achieve your quarterly goals. In many cases, involvement from a product manager in a key sales deal can be the difference between missing your product's number and hitting it.

Develop a relationship with your colleagues in both sales and finance who are responsible for tracking your product, and they can help you with analysis that will enable you to help grow your business in powerful ways.

Industry Data

This includes any data not specific to your own business that provides insight into the direction of the industry in which you compete. We discussed several of these sources in Chapter 6. It includes your Total Addressable Market (TAM), industry studies and analysis, competitive intelligence, and more. When seeking to bolster your argument that a certain customer problem is worth solving, having access to a good TAM or a survey of C-suite leaders suggesting that 78% of them feel like the problem you are advocating for is business critical, can make all the difference.

Testing

Nearly every product management thought leader outside of enterprise software, and many within enterprise software, advocate for heavy data-driven testing. Testing can take many forms, and some of them are highly advisable in any product management context. Here we will focus on a few ways in which enterprise software is different from our consumer software friends.

Let's get heretical. Usually, we are huge fans of A/B (or "split") testing. The theory, which is sound, is that by trying out an idea (a new UI design, a new feature, etc.) in a statistically valid test, you can learn with confidence (at scale) which option users prefer, and then roll that preference out to all customers. If you're not sure how to improve your customer experience (and who is ever really sure?), run a test and let the results speak for themselves. And in consumer software, and in limited ways in enterprise software, A/B testing is a powerful (and often cheap) way to ensure that what you ship drives the KPI you intended it to drive.

Here is the rub, though: when big companies pay you millions of dollars for software, the last thing they want is major, unannounced, and unexpected changes to the product. This is even more true for business-critical applications like the ones you probably work on. If one of these goes down, or changes in unexpected ways, it's not just annoying — it's lost business for your customer. The stakes of failure are potentially huge, and not just because customers expect the product they paid for to always be available.

The reason A/B testing is great for most websites and consumer applications but sometimes risky within enterprise software products is that enterprise users (i.e., those "pay[ing] you millions of dollars for software") do not like to be jerked around when it comes to their user experience (UX). Nobody likes it when things suddenly change on them, but the relationship is quite different when your customers are massive organizations rather than individual customers. You will likely find in enterprise software that the companies you sell to—which might have hundreds or thousands of users consuming your product—insist on *months* of advance notice when there are changes to the UX. Your power users then train their (often very large) organizations on how to use your tool. When you change it on them in the name of testing, they are often left hampered in their ability to navigate to the key workflows that provide value.

From a testing perspective, the reason this doesn't work is that test participants aren't supposed to know that they are in a test. Let's consider a new user of your tool. We'll call him Charles, and he's just gotten access to your product. You want Charles to have a positive experience because he is part of a company paying a lot of money for your software, and the tasks he completes with your product may well contribute to the customer's view of their return on investment with your company. To become familiar, Charles watches a few training videos recorded by your education team. He reads a bit of documentation, as well.

Now Charles logs in. But wait! This doesn't look quite like the videos. Where is the feature he read about in the documentation? This doesn't look right! Charles is confused and frustrated. To make matters weirder, Charles asks a friend, Amanda, at a different company why the product doesn't match the documentation or videos. But Amanda sends Charles a screenshot: to her, the product looks exactly like what is documented.

What happened here? Charles ended up in a certain test group, and thus received a different variant of the UX than other users. Amanda was in the control group, which got the traditional UX. There could be multiple test groups, each with a different experience. Some of these experiences might indeed be better than the control, and users might be able to find their way painlessly. But many of them won't be, and frustration will mount. Nobody is actually going to quit Facebook because their News Feed looks different than someone else's News Feed (and even if a few people did, it wouldn't put a dent in Facebook's overall growth), but they often will abandon your software, putting revenue at risk for you.

Why not simply create separate training, documentation, and so on for each test variant? This certainly would help, but show me the software company that has enough resources and coordination to support creating multiple versions of all supporting materials and I'll show you the Lost City of Atlantis.

(Another potential issue here is sales. The last thing you want during a demonstration to a prospect is an unfamiliar experience that your sales team doesn't know how to navigate. You can mitigate this by excluding demonstration accounts or demonstration environments from tests, but this is still risky at best; when the customer buys, what they see might not match what was demonstrated.)

We will stop being heretical for a moment and offer some additional clarity: there is certainly a place for testing in enterprise software. However, if you come from a consumer software background and were used to running multiple tests with major user workflow implications each month, you might find it difficult to develop a similar program in the enterprise world. We recommend starting small, testing with statistically significant but small groups of users, and focusing on minor changes to your product that won't send a user or a sales rep diving for the documentation when things suddenly look different than they used to.

Another good approach to "testing at scale" in enterprise software is to let users "opt in" to testing via a "public beta" or other "labs" mechanism, so that those who feel comfortable finding their way around potentially unfamiliar features or workflows know what they are getting into. This is not strictly A/B testing, but is a good way to gather feedback on changes to your product before they go live so that you can ensure your improvements will have their intended effect and solve problems for your customers.

If the theory of testing is sound, but the practice as often evangelized in product management does not work the same way in enterprise software as it does in consumer software, what is a product manager to do? For major product changes, we recommend focusing on extensive *user testing*: using quickly built prototypes to allow users to explore your concept as they would any other part of your product, and then iterating quickly on those prototypes, followed by more testing—lather, rinse, repeat.

Prototyping involves UX designers and/or developers creating a light version of the proposed change. This can be in a development or beta environment, or can even (in the case of UI updates on the web) be a JavaScript bookmarklet that appears to change the layout and CSS in your product for a user, without actually making any underlying code changes.

User testing involves working with a small group of users, typically of varying experience levels and possibly in different industries, to see how well they understand the prototype. These tests should likely be run by your UX design team, with product management and development supporting. The test administrator should give the subjects a series of tasks to complete given the proposed new experience. Their ability (or inability) to adapt and intuit their way through the new experience while talking you through their thought processes will yield valuable insights.

In some cases, you might even be able to "A/B test an experience in production" by explicitly inviting users to join a test. This would drop them into a variant on their normal experience, and they can provide feedback via a form. Your most enthusiastic power-users will often be eager to oblige.

Prototyping and user testing early in the software development process will not only ensure that the experience you're building is easily understood, but it will allow all of the supporting materials (documentation, training, etc.) to be created around a polished final product, so that there is harmony across all of the "channels" through which users master your product.

In sum: our advice is to test away, but do it in a way that's right for your enterprise users. Involve your customers in testing prior to (and during) product development, but don't risk alienating your users, whose experiences pave the road to your future earnings. Don't give them unexpected product variants on an experience on which their business relies.

Summary

Enterprise product managers often have copious opportunities to use data to better understand their customers, and this data sheds quantitative light on other research you will do with customers. We recommend making analysis part of your product management process wherever possible. We also advise a bit of caution around A/B testing, but you will have plenty of opportunities to do both split testing as well as user testing involving prototypes. All of these methods will help you to ensure that you are solving the right problems in the right ways.

PM Profile

Name: Kimen Warner

Title: Group Product Manager, Adobe

How did you get started in product management?

I majored in product design in college, which was essentially product management combined with mechanical engineering, so I learned how to actually build physical products. I knew I wanted to do software product management so I could iterate quickly (and honestly not spend time in factories!), but it's almost impossible to find a product management role directly out of college. Instead, I went into consulting and then joined a startup as a client-facing technical consultant where I learned the product. I then leveraged that product knowledge to move into the product management team. That startup got bought twice, and I'm still doing product management for it!

Are there particular backgrounds you think are especially valuable for enterprise product managers?

I think having some technical knowledge is crucial. Nobody expects you to code, but knowing what's hard and what's easy, and how a design might get implemented makes the entire process faster and builds trust between you and your development team. The other piece is customer relationship experience. Enterprise product management means heavy customer engagement—from pre-sales to managing unhappy customers after they've experienced a product issue. Diplomacy and confidence in those conversations are essential for success.

What have been the most difficult parts of your job to learn?

Feature prioritization is tough. There are always many more things we want to build than we can, and the argument for each one can be extremely compelling. Every rubric I've tried has its faults, and it only gets harder when promises have been made in sales cycles, by executives, and others. It's a constant struggle to ensure we're doing what's right by our customers, our bottom line, and our thirst for innovation.

What would you do in tech if you weren't in product management?

This one's tough! I spend more time thinking about how I can stay close to building products while becoming more senior in product management. I'd probably be most interested in a "solutions architect" style role, where I'd work with customers to try out new and complicated ideas using a company's technology. It would let me "geek out" a bit and kick the tires on new ideas to solve problems. But honestly, my heart is really in product management.

Industry Knowledge

Product managers cannot help design the future of the product without knowing where their industry is currently, as well as where it is headed. Fortunately, you will have many tools at your disposal to gain this knowledge: most important, insights gleaned from working directly with your customers in a number of ways.

Although we have not talked much about them so far in this book, we will admit that there are some ways in which product management for enterprise software might arguably be *easier* than in consumer technology. We're going to discuss one such example in this chapter: gaining an understanding of your target market. Of the three types of knowledge discussed in this chapter, industry/market knowledge is the one that both new and experienced product managers in an industry will need to continually strive to obtain, especially given the pace of technological change in most industries.

Consumer technology has its addressable segments, of course. But even so, at large scale there is a tremendous diversity in the kinds of use cases normal technology consumers have. In enterprise software this is still a problem, but is somewhat easier to overcome. Businesses in the same industry tend to share many characteristics in common: how they sell; what types of products they sell; how they source, produce, and distribute those goods; and so forth. Knowing a lot about how an industry works gives an enterprise product manager very valuable insight into what types of problems people in that industry face.

In mature industries, there can be a lot of noise blurring the signal you want to capture. Sometimes, the industry old-timers who think they know everything are plain wrong about what's coming ahead. Or, perhaps the trade press is institutionally invested in a narrative that's out of sync with what problems really face the industry. (Just, you know, some wild guesses!) With nascent industries and markets, you will have much less definitive data about what's going on, especially in the public purview, making direct customer input all the more inimitable.

But perhaps the best reason to develop your industry knowledge is—you knew we'd say it again—to better develop empathy with your customers. The reality is that even the best product managers will only ever likely meet a small sample of their end users *or* customers, particularly if your product is a large, mature one. You can extrapolate a lot of valuable lessons from that sample, to be sure, but it's difficult to be sure you have minimized all the bias you can from it. Developing a keen sense of the industry you sell into is the best way we know how to deeply understand the personas and roles of the folks you care about. The more you understand about their industry, the better you can grasp why your customers do the things they do and begin to recognize your own blind spots as an outsider. Only after you have a good sense of those can you truly begin to imagine how to build tools to innovate forward in that industry.

Let's begin with four specific topics that should be applicable regardless of industry maturity. There are many others, but these will serve as a strong foundation for building your market knowledge.

The Basics of Your Market

Your Total Addressable Market (TAM). Your product has a TAM, but even a specific feature might enable you to compete in a new or broadened market, with a separate TAM. If your company is reasonably mature, it will have a developed corporate strategy function which researches and updates TAM calculations for any market where you already compete, and it should be eager to help develop new TAMs to support the growth that comes with entering new markets successfully. If none of this is in place at your company, you can begin by estimating. Note that you can often get TAMs from firms like IDC or other industry analysts, but in our experience software companies' own estimates are often far superior, in part because they give you the opportunity to define a market in your own terms rather than someone else's often too-generic terms. (In fact, we often find that estimates from "operators" in these markets are usually far more realistic than those of bankers or analysts.) Look at your competitors' financial statements or funding, and work backward to a guess at revenue. Add that up across all major competitors, and you are on your way. Although it is best to have an "approved" TAM calculation, the essence of the thing is most important: be deliberate about understanding your market share and your total opportunity.

The great thing about a TAM estimate is that it provides context to both the value to your customers of solving a problem *and* the value to your own company of solving that problem. Suppose that you have a $500 million TAM and you think you can realistically capture 10% of it within the next two years by taking on a certain project. The $500 million number gives you a way to compare the size of the customer problem with other problems you could solve; in other words, knowing that the market can support $500 million of spend to solve a problem (or set of problems) gives you a way to gauge which problems are most worth solving. If you could reasonably pursue 10% of a $500 million market or 10% of a $2 billion market, of course you know which you would choose. Beyond that, though, the estimate that you think you can capture 10% of a $500 million market tells you that the opportunity for your company is worth $50 million.

Closely related to TAM (and usually included in any TAM data that your corporate strategy team might offer you) is Compound Annual Growth Rate (CAGR), which conveys the rate of growth for a market. We will not do an in-depth exposition of CAGR (you can easily Google it), but simply put, it helps you to understand how fast the set of customer problems encompassed by a TAM is growing in importance and value to your potential customers. Consider the two opportunities presented in Table 7-1.

Table 7-1. An example TAM and CAGR illustration

	Market share in five years	TAM	CAGR (next five years)
Market 1	30%	$500 million	24.5%
Market 2	25%	$750 million	7.5%

At first glance, 30% of $500 million is $150 million, whereas 25% of $750 million is $188 million, so perhaps you would rather attack the problem set represented by Market 2. But CAGR adds another layer. Realistically, actually capturing additional market share by shipping new products and features takes time. This is even more true in enterprise software with direct sales models, where you are bound by sales and renewal cycles; in our world, most customers are not simply rolling up to an ecommerce form and entering a credit card to buy your software. Five years from now, 30% of Market 1, with its 24.5% CAGR, becomes a $360 million opportunity, whereas 25% of Market 2 at 7.5% compound annual growth is a $250 million opportunity. Remember not to take TAM at face value, but rather factor CAGR into how you think about selecting your investments, because you are not competing in static markets.

Your market drivers. What are the major trends in the market that are affecting how people think about your product and your industry? Is the explosive growth of mobile connectivity changing how consumers interact with the brands you serve? Do advancements in CRM technology mean that more brands are presenting personalized offers to customers? Is the development of better in-memory databases changing cost profiles for your customers? Are there mega-competitors in your customers' industry that all others are forced to compete against? Learn what these trends are and have an informed guess at what they mean for the future of your product.

Where does this knowledge come from? To some extent, it simply requires being a student of technology, and not just in your own industry. Market drivers often seem to have little or nothing to do with your product. Like the analogy that "it is the airplane you *don't* see that will shoot you down," technology trends that seem totally unrelated to your product can end up turning your entire industry on its head. So, you need to be aware of the world around you and observe how technology is changing other markets and lives in both the business and consumer world.

Both of us have worked as product managers in the web analytics industry. In 2009, web analytics was humming along and taking over the world as the enterprise at large was really just beginning to grasp the value of understanding their consumers' web behaviors. It would have been easy for anyone in this field to have tunnel vision and double down on the web. But an astute product manager in this industry would have seen the wild success of the iPhone and, in particular (hence the choice of the year 2009 for this example), the App Store and immediately realized that the market was about to be driven by a complete reorientation of how brands and consumers interact, and indeed how consumers think about technology at a very fundamental level. The introduction of the App Store (and Google Play) did not directly, on the face of it, have anything to do with web analytics, but at the same time, it had *everything* to do with web analytics. The web analytics market was about to be entirely superseded by a "digital analytics" market that treated web and mobile app as parts of the same customer journey. Awareness of both the growth of mobile *and its potential impact on the web analytics industry* as *market drivers* would have required a product manager to be paying attention to technology outside of her own world.

If you are successful in recognizing the most relevant market drivers, you can often find emerging markets or emerging personae in existing markets (such as the "mobile product manager" shortly after 2009), both of which can represent opportunity for you as a product manager.

The most common problems your customers and prospects face. What is preventing your customers from growing as fast as they would like, and how can your company help? If you spend enough time talking to customers, you will see patterns begin to emerge. They might not all use the same words, but you will recognize that they are really asking about the same things. (As perhaps an extreme example, we have had experiences in which five consecutive meetings at industry conferences all centered around the same general problem.) Even though there are a number of ways to learn about the problems your customers and prospects face, we do not believe that there is a worthy substitute for spending time actually talking to your customers about what is on their mind. Getting customers to open up to you about what really worries them is a difficult task.

Your competitive landscape. First, a word of warning: too many product managers become obsessed with the competition, usually unintentionally, without knowing that it is happening. Through the process of competition, they end up relying on the market leader's product strategy to define their own, or believing that knowing their competitor inside and out will allow them to fend off threats. Neither of these approaches works in the long term. That said, it is valuable to know who you compete with, why they sometimes win, and what your relative strengths are. This is where a good ol' fashioned "SWOT analysis"[1] is worthwhile. From a market knowledge perspective, this is important because your sales team will need it, and because you can often spot gaping holes in your competitive set offerings where there is a problem that no one is solving.

Remember that the purpose of market knowledge is to allow you to set and manage the vision for a product or feature. Anything that helps you walk into a room full of executives and explain why customers need whatever you're proposing is good. We'll get into how you should lead that conversation later in the book.

Learning Your Market

The process to obtain market knowledge, although often difficult for newcomers to an industry, is also fairly straightforward. As with all types of knowledge, this takes time to master, but the following resources can help accelerate your expertise.

CUSTOMER MEETINGS

As mentioned in the previous section, nothing can replace this. Most people like to talk about what they do and why they do it, especially if you've earned their

---

1 Strengths, Weaknesses, Opportunities, Threats: *http://bit.ly/2EiIG3p.*

trust. We have usually found customers willing to be interviewed and, more often than not, willing to share their views of their own companies and the challenges they are facing. On one team that we observed, product managers were required to conduct at least six in-person customer meetings each quarter, which forced otherwise busy product leaders out of the office so they could learn about market and customer problems. Even if you cannot travel extensively, phone conversations are a great start. It can be helpful to work up a basic mutual Non-Disclosure Agreement prior to these conversations to enhance your customer's trust in you and also to allow you to be a bit more forthcoming with your own product strategy than you would be in public. Remember: you're trying to get to the essence of what keeps your customer up at night. That is where you will find willingness to pay for software.

Note

This advice is also applicable to meetings with noncustomers. Although much of your time will likely be spent with customers because they have knowledge of your current product offering to serve as context when discussing their business objectives and problems, prospects, former customers, and companies who chose your competitor over you all provide worthwhile learning experiences that will also give you deeper insight into customer problems worth solving.

Whenever possible, have an engineer or two and a UX designer join you for these meetings. Engineers are great at solving problems. They are where much of your best innovation will come from. Connecting them directly to the customer gives them an (all-too-rare, unfortunately) opportunity to develop empathy on a first-hand basis. Rather than having you send them a bunch of notes that you took, which can read as dry and impersonal, they have an opportunity to meet a real human being, with real motivations and real challenges. This will hammer home the frustration inherent in the customer problem and generally produce better software solutions at the end of the development process. We cannot stress enough how valuable it is to involve engineers and designers in these customer meetings.

There are a variety of types of customer interviews, and all of them are instructive in different ways; when you employ each kind will depend on a variety of factors such as where you are in the development process for a given project, your customer's maturity, and more. Let's focus briefly on a few that are particularly helpful for understanding *customer problems*.

Discovery interviews

A discovery interview is probably the most common type of customer interview driven by the product team. It involves sitting with one or multiple individuals on site at a customer's office and having some form of open dialog. Both/either engineering and UX can be involved, too (in-person or remotely), or at least another product manager so that you have multiple ears on a discussion. The goal of a discovery interview is to better understand the challenges facing your customer that are preventing them from growing. As you meet with many customers, patterns will emerge. You will make connections between one company's inability to determine ROI on marketing spend and another company's questions about the consumer journey that will crystallize in your mind a direction for your product or feature.

The purpose of discovery interviews is *not* to have the customer tell you what features they want to see in your product or which companies you should acquire. Customers love to try to innovate for you, and of course it is fine to humor them. Every now and then you come across a truly great idea, and it's good to encourage your customers to think as technologists, but they do not have the perspective that you have. They are not the ones talking to dozens of other companies about their problems and identifying patterns that point to real opportunities. An individual customer has a unique set of problems, and wants those problems solved *today*. Your job is to figure out what problems they are going to have tomorrow, and solve those.

Mark Cuban of *Shark Tank* fame (among other notable ventures) described a software company that once had a best-in-class product:

> *Then it made a fatal mistake. It asked its customers what features they wanted to see in the product, and they delivered on those features. Entrepreneurs [and also product managers!] need to be reminded that it's not the job of their customers to know what they don't. In other words, your customers have a tough enough time doing their jobs. They don't spend time trying to reinvent their industries or how their jobs are performed. Sure, every now and then you come across an exception. But you can't bet the company on your finding that person among your customers. Your customers can tell you the things that are broken and how they want to be made happy. Listen to them. Make them happy. But don't rely on them to create the future roadmap for your product or service. That's your job.*

Because you are trying to invent the future, not just deliver on what customers want *today*, asking them what they want, or whether they would buy Feature X, is usually not particularly effective. But digging into their business needs with genuine curiosity about the way things are done, why they are done that way, and where there are gaps preventing growth, is tremendously effective. Sometimes, the product itself hardly comes up in these discussions, which may counterintuitively mean that you're on the right track.

In your time as a product manager, you will become an expert interviewer. You might soon find yourself critiquing the beat writer asking questions of players after a professional basketball game, or the talking heads on your favorite news outlet. Conducting a good customer interview is not all that different; you want to ask good, thought-provoking questions that are relevant to the topic at hand and be able to truly listen and dig into follow-up questions based on the individual's response. Here are a few other keys to conducting a good discovery interview:

- Talk to a variety of personas, including the buyer, your primary/secondary/ tertiary users, and more. Remember, you're trying to get a feel for how a company is growing or not growing, and many roles factor into that story. If you are working on an email marketing product, don't just talk to the email marketers. Talk to the product managers who run the ecommerce funnel. Get their take on what the company's biggest marketing challenges are, too. This can be difficult, because individuals unfamiliar with your product might be less willing to make time for you, but with patience and clear instruction, some of your customers will introduce you to people who will give you a completely different (and very valuable) perspective.

- Bring a series of general questions in addition to any that you might have about a particular product area or problem set. Here is a set of questions that you might bring to every discovery interview if you were working on a business intelligence product:

    — What do you think is unique about the way your company does business intelligence? What is unique about the way your company thinks about data?

    — What integrations are most important to the growth of your business?

— What is a typical customer journey, and how do you see your company influencing that journey with data and insights today? Next year?

— How is your organization—and the organization around you—structured?

— How do your interactions with various other teams take place? Channel? Frequency? Tone and types of requests?

— How has ___ changed here over the past two years?

— What are the top three things preventing your business from growing?

— Why is your business intelligence organization structured the way it is?

— What is your strategy for going forward?

— What do you hire business intelligence to do?

— How do your executives view/think about business intelligence?

— Tell me about a time that analysis (in any context) made you look like a rock star.

— Tell me about a time that our product delighted you.

— What is an example of a product that delights you to use?

— What are you trying to do with your digital data and customer data that is really difficult to do today? Why is it important?

— Who are your top three to five nonanalyst roles who rely on business intelligence/data? What are their daily lives like?

— How does your organization set goals? How do you forecast results?

— How are data and analytics a competitive advantage for you versus your competition? In what ways do you think your competition uses business intelligence better than you?

— What are the top three business problems that your company is working on solving with data today that you weren't six months ago?

Notice that these are open-ended questions. Avoid yes/no questions because they won't help you dig deep into an individual's motivations or underlying needs.

- If you ask a customer what challenges the business as a whole is facing, they might or might not get right to the heart of the matter, but this is why (no pun intended) we love the "Five Whys" root-cause analysis method of interviewing. The theory is that within (roughly) five iterations of asking a person "why?" you can get to the root cause behind the initial problem described. You ask what problems are facing the business, and they give an answer. You ask why that's a problem. They answer. You ask why that's a problem. They answer. Lather, rinse, repeat. By the fifth "why," you should be at a root cause of a problem that motivates the users you're interviewing to action in a big way. That should either be the customer problem or very close to it. For example:

  Q: Why are people at your company not using the dashboards your team builds?

  A: Because the dashboards don't really answer their questions, I suppose.

  Q: Why don't the dashboards answer their questions?

  A: Because we can't visualize our sales and manufacturing data in a way that makes sense to leadership, and without leadership's interest it isn't worth anyone's time.

  Q: Why can't you visualize data in a way that makes sense to leadership?

  A: Because they think in terms of a supply chain, but our software options for reporting back to them don't naturally think that way.

  Q: Why do they think in terms of a supply chain?

  A: Because that is really where our industry's bread is buttered. Reducing supply chain friction is the best way for companies like ours to reduce costs, boost sales, and stay competitive.

In this example (which actually only took four "whys" but could easily have included a fifth that dug into why supply chain management is the key to growth in that industry), the product manager asking questions established the core of business growth rather than staying focused on the software. As a product manager, this is exactly what you want. There are countless ways to improve your product or your customer experience, but without a fundamental understanding of how these improvements might affect your customers' ability to grow, you are likely grasping at straws. Note that the product manager in this example could have broken off after the first "why" and asked, "How can we improve our dashboards?" That is a fairly common way that many product managers might approach a customer interview. The interviewer might have received a response such as, "We need better visualizations." The customer might have even said, "We need a way to visualize our supply chain." But by performing a root-cause analysis and digging deeper into the problem, you give yourself the foundation needed to come up with solutions that your customer can't even imagine yet.

Observation

The oft-repeated maxim that "customers often say one thing but *do* something else" means that interviews alone won't always reveal everything that is vexing your customer. Thus, whenever possible consider including *observation* in your customer interaction repertoire. Observation can sometimes be the most difficult type of visit to pull off depending on how end users consume your product, but often the most valuable. Observation, sometimes referred to as *ethnographic research*, means literally sitting with end users of your product to see how they interact with it. Unlike discovery interviews, the content of this type of visit isn't controlled by your line of questioning; it is determined by the end user's day-to-day role. Job shadowing can also fit into this category and might add even more context for how your product as well as its potential outputs are actually used in your customers' organizations.

Ideally, when doing ethnographic research, you would ask questions only when you don't understand something that the subject of your observation did. Other than that, you let them work and notice where they run into roadblocks, where workflows fall down, where they need to work with others, and so on.

Compared to discovery interviews, where your goal is to learn as much as you can about customer problems by listening and questioning, observation is more likely to reveal ways that you can improve your customer or user experience.

Quarterly Business Reviews

Quarterly Business Reviews (QBRs) have many names, but whatever your organization calls them, in enterprise software they represent an opportunity for account teams to check in with customers and sow the seeds of the next deal. Almost always, you will join these by invitation, usually to present the product roadmap or discuss recent updates to your product. As with any request to present rather than to listen, your challenge as a product manager is to make the most of the situation.

You will need to work with your management to find the right amount of involvement in QBRs; left unchecked, these can end up consuming too much of your time and distracting you from building and shipping software. You don't always need to be involved with QBRs when asked.

Although you do not want to become a machine, regurgitating your product roadmap on command, there are a number of benefits to joining QBRs (and to presenting your product roadmap in customer meetings generally). Of course, there is the opportunity to ask probing questions as you share what your team is building, and to turn those questions into customer problem discovery sessions. But even without that, presenting the roadmap helps you to get your name out there with your sales team and with customers, which helps build your brand both internally and externally. Additionally, as you adapt your explanation of roadmap features and use cases to a specific customer in a given industry, you will find yourself thinking up new use cases and requirements that you might have missed previously. Finally, it helps you internalize your product strategy, which will not only help you obsess about how you solve customer problems related to that strategy, but will also help you explain your product vision and strategy to internal stakeholders as you work to convince them of your ideas.

User testing and prototyping

The last type of customer meeting that we will discuss in this chapter (although there are many more ways that you will interact with customers) is user testing and prototyping. Although the other three types of meetings we discussed might or might not be tied to a specific project or feature idea, user testing and prototyping always are. There have been dozens of books written on the topic of prototyping, and there are myriad software companies that serve this particular need quite well, so we will give it only a cursory attention here.

When prototyping, you are showing to an end user either a product design (mock-ups) or, ideally, an interactive, working version of something you intend to build to see how they engage with it and get their feedback. A pillar of product development in small, "agile" startups, prototyping tends to fall by the wayside in enterprise software, which is unfortunate. Whatever your current development method, work with your engineering counterparts to add prototyping to the process if you can. The feedback you get from frequent, rapid prototyping helps ensure that the feature you ship actually solves the customer problem that you set out to solve in the first place. Without it, you're only guessing, and even the most astute product manager will guess wrong half the time (or more).

When we say "rapid" prototyping, we mean that the more rounds of proto-typing—meeting with a customer, testing your prototype, and implementing their feedback into the prototype and product requirements—you can do, the better.

GETTING CUSTOMER MEETINGS

Getting customer meetings scheduled and planned also happens in a variety of ways. If you've been working on your product or in your industry for a while, you might have contacts at various brands, with whom you can work directly to schedule any of the aforementioned types of customer meetings (except QBRs, which are typically led by account teams). More on this in a moment.

If not, don't worry; there are two great ways for you to find customers to meet. And only one of them comes with a warning attached.

First, use data. If you have a hypothesis about a customer problem and want to validate it, or if you have an area of ownership where you are looking to help your product take its next big step, various data can help you find customers who might be perfect targets for discussion or observation. Even something as simple as "customers who aren't using our recently released features" or "one-billion-dollar-GMV (Gross Merchandise Value) retailers located in the US" can narrow down your search and help you identify good candidates to meet.

Second, use your field organization. As a product manager, you have purview over your entire customer base, which means that you often get to know a lot of customers, but not as deeply as good account teams do. When you're ready to dig into customer problems, talk to your customer success or sales teams to find out who might be open to brokering a conversation. In many cases, these groups are thrilled to help you connect with their customers, because attention from the product team shows how much everyone cares about the customer's business. In some organizations, by contrast, account teams can be cagey or protective of access to customers, either due to the risk of overwhelming them or because they don't understand product's needs for customer input. In these cases, explaining your approach to sales organization leaders first can open a lot of doors.

A few warnings here, however. Beware of allowing the sales team to sit in on these customer interviews. Often, in subtle but meaningful ways, the presence of a team whose expressed purpose is to sell more software will change the way a customer responds to a question. You need the customer to have complete freedom to say anything, without fear that it will come back to hurt them in a future negotiation or sales pitch. Typically, product managers carry credibility in this regard which helps customers open up to them. You will need to be polite but firm when scheduling a customer meeting through an account team: you are asking for a favor, but they are not invited to the dance. (A compromise that often works is to have a portion of the meeting during which sales is included, and a part when they step away, or—even better—to meet alone with the customer and then have sales take everyone to dinner afterward!)

One of the common questions that a product manager faces is which set of customers from which to prioritize learning. This is a challenge largely unique to B2B software, because implementations, use cases, and contracts often vary so widely. Do you focus on the high end of your market, your biggest and loudest customers? Do you sample from each segment of your customer base? There isn't a single answer here. It depends on a variety of factors: the maturity of your product, your growth strategy (Are you trying to go upmarket? Or downmarket to hedge against a disruptive competitor? Are you expanding into an adjacent market? etc.), and many other things. However, it is something to be deliberate about. For the purposes of gathering knowledge, you will not always treat every customer equally, and that's okay.

In each type of customer meeting, make sure you are clear ahead of time about who you need to meet. If you know a bit about the customer problem you are hoping to solve, you should be able to hand-pick the roles that will be able to tell you about that problem most effectively. If not, meeting with executives or senior resources who have good visibility into the challenges facing the organization is a good place to start.

Trade press

If you compete in a big, mature industry, there are likely to be people writing about it already. Certainly, big industry segments like retail, travel/hospitality, healthcare, marketing, finance, nonprofit, and others have any number of trade magazines, blogs, newsletters, and other outlets that do nothing but study their own industry. It may well be impossible for you to consume all of what's written about your industry (it certainly is for us), and you certainly shouldn't try. As you become more familiar with the dynamics of the industry, you'll be able to skim the major outlets and get a sense for the broad themes they're focusing on.

Analyst/research reports

Although they are far from perfect, and occasionally laughably reductive, industry analyst firms serve a worthwhile purpose in the market knowledge sector: they talk to *everyone*. They talk to your customers. They talk to your competitors. They talk to your competitors' customers. And they talk to you. The reports they put out are the result of that research, and often point out market problems within the context of existing offerings. These firms also frequently publish surveys that help to quantify market problems. For example, knowing that 74% of COOs expect to increase spending to solve a particular problem in the next 18 months gives you part of the evidence you need to go make the case to build a certain product. Analyst reports usually are not cheap, but they often provide some information free of charge and can be a good use of budget if you believe they contain an insight that will substantially help your argument. Larger organizations might have a subscription to analyst reports, so check with your analyst relations team to see what access you might already have. Be careful not to over-rely on what the analysts are saying, though. They're a good input to have, but the best product managers are already their own analysts.

Consume everything

This might go without saying, but take advantage of publicly available channels: blogs, forums, podcasts, and so on. You will find market knowledge in places you might not expect. A forum posting asking for help doing a certain pragmatic task in your product can give clues about the underlying problem they are trying to solve. The authors of this content might also make great interview subjects to help you delve deeper into their challenges.

Competitors

Efforts as simple as reading your competitors' blogs will help you understand how they view the market and what their product strategy is. You can decide whether you want to try to beat them at their own game, or take a complementary approach by going after a different niche or set of use cases. Reading their blogs also helps you to understand their marketing, which is how most tech companies telegraph their product roadmaps. This will tell you how they view the market, as well as their buyers and their go-to-market strategy, all of which influence their product. If their view of the market seems significantly different than yours, take note—are they wrong? Or might you be?

Win/loss analysis

One of the most commonly overlooked sources of market wisdom is win/loss analysis, which you should help conduct in partnership with the sales team. Work with sales reps after a deal to understand why the customer (or noncustomer) made their buying decision: what the most important factors were, what your company could have done differently to win the deal, and so on. To be clear, understanding why you *lost* a deal is typically more valuable to a product manager than understanding why you *won* a deal. Both help convey market knowledge, but the former gives a much clearer signal about what is lacking in your current offering that will help you compete better in the future. At the same time, don't slip into "me too" or "check the box" mode; after you know what your product lacks, figure out how to make it solve the problem in a new way, especially when that anticipates future trends in your market. There are even third-party firms that specialize now in helping software vendors perform win/loss analysis (and we can help recommend some great ones if you're interested).

Surveys

We have found that periodically surveying our own customer bases can provide unique market knowledge that, like the surveys conducted by analyst firms, gives a quantitative element to your market knowledge. One clear consideration here is selection bias, so make sure to ask questions that won't be influenced (too much) by the fact that respondents are already your users. If you have the means to do true random sampling or to survey *noncustomers* as well as customers, we strongly advise it. If not, recognize and be open about the fact that the survey results can be colored by familiarity with, and preference for, your company and its products.

Customer Advisory Boards

Last but certainly not least, establish (if it does not already exist) a small group of customers to serve on an advisory board for your product. There are a variety of ways to organize and conduct a Customer Advisory Boards (CAB), but ensure that you fill it with individuals who can impart real *market knowledge*. These might well be the executive "buyers" of your product, but not necessarily. They should be people who know their company and your industry well enough that they can help point out holes in your product strategy, or warn you about upcoming trends to be aware of. If you are fortunate, they might even be people who neither buy nor use your product, but who can share with you and the other members of the CAB a unique perspective on an adjacent market or persona.

It can be tempting to staff your CAB with "power users" of your product, but we have found that such individuals might operate at a more tactical level and lack insight into topics such as budgetary decisions, corporate strategy, and so on. Remember, you want people who will be able to look at their own companies and answer the question, "What keeps our CEO awake at night?" You will also want people who will be brutally honest, but who also want to see your company and your product succeed. You'll want to meet with your CAB regularly, but cadence and medium (in-person, video conference, etc.) will vary based on your needs.

Industry conferences

Here, we don't mean tech events (though those can be useful in other ways). If your target industry holds conferences periodically that focus on major trends going on within it...well, it's a no-brainer that you might want to attend. Even at the mega-conferences, there is much to be learned simply by attending speaking sessions, panels, and perusing the vendor showcase floor. The smaller conferences can be uniquely valuable, too. The types of attendees at either end of the scale tend to be different, which can result in refreshingly varied perspectives. When members of one of our teams attend such a conference, it's standard practice to write up a trip report and share it with the rest of the team when they return, explaining what they saw and what the rest of the team might be able to learn from it.

Social media

You knew we were getting here, right? Both @BenjaminGaines (*http://twit ter.com/benjamingaines*) and @BlairReeves (*https://twitter.com/blairreeves*) are enthusiastic Twitter users. In our industry (marketing technology), social media is a major source of both market sentiment, competitive intelligence, and, not infrequently, user feedback. Almost any industry has some community of users on social media (not necessarily just Twitter, but there especially), and it's a smart move for product managers to have a presence in those online communities, as well. We both believe that Twitter as well as blogging has been absolutely key to our careers in enterprise product management. Your mileage may vary, and, of course, don't pour too much of your time into this area; but as a source of industry knowledge, social media can be a major resource.

After the Customer Interview

Part of the value of spending time with customers comes from simply learning to think like they think and developing empathy by seeking to understand them. There is a lot of value in simply spending time with customers so that you can represent them in the proverbial room with your developers and designers, ensuring a better product. But much of your more pragmatic learning from customers will come when you can bounce your notes and impressions off of colleagues. If you are part of a product management team, hopefully your teammates have also been out and about, holding interviews and learning about customer problems. By sharing and analyzing your experiences together, you will often reach new conclusions that hadn't been apparent to you previously.

There are a variety of ways to do this, and there is no one recipe that works in every company, team, product maturity, and so on. Here are some common approaches that may be effective for you:

- Hold a team meeting (weekly, biweekly, or monthly) for the purpose of reviewing customer interactions. Talk with your colleagues about whom you met, why you chose them, and what you hoped to learn. Further, discuss how the conversation progressed, key insights you picked up on, and any action items coming out of the meeting. This should be less a presentation and more a discussion, where you invite your colleagues to probe deeper on certain topics that you share from your interview. You can end up veering off in other directions not related to your customer interview, and that's okay. The interview becomes a data point in a rich sea of information that can help you identify patterns even outside of your own product focus.

- Share your interview notes on an internal wiki page or similar widely available internal site, and provide the link to your teammates along with *specific* requests for input or feedback. (One advantage of this approach is that it scales almost infinitely; discussion becomes ineffective with more than a handful of people in the room, but you can send the wiki page to hundreds of product managers, developers, designers, marketers, and so on. When these roles can't be in the room with the customer, making sure they see what you learned is the next best thing.) The key to getting value from this approach is to ask your colleagues to review specific points in the document and share their thoughts. Because product managers are typically pulled in a thousand directions, a link to a set of notes will often go ignored unless there are three to five clear and specific things to which you are asking colleagues to respond. For example, you might ask, "In these notes you will see that XYZ Corp said they don't have budget to spend on both software and the consulting required to get real value out of the software. I would love your help reviewing what they said about this. In your experience, is that a challenge we simply have to accept, or is there something you could see us doing in our software to limit the need for services?" Give your teammates a few questions like that, and watch the ideas (and patterns of ideas!) begin to flow.

- Distill the customer problem into a written story or use case, and distribute that to teammates for feedback. This carries with it the advantage of conciseness, and turns your lessons into something concrete that looks like justification for doing something. This can also be the most effective way to produce empathy for your customer in others. You can ask your colleagues to review these customer problems and speak to whether these problems would be worth addressing in the product at the expense of other things you could do. We have seen product managers turn these stories into actual recorded videos, with "actors" (usually the product manager and colleagues) playing the customer, who is trying (and failing) to get something done using software. When done well, this approach to sharing what you learned in a series of interviews can be extremely effective in deepening awareness of, and empathy for, customer problems.

Whatever approach you take, don't let the knowledge that comes from your customer interviews stop with you. The entire product team will be more successful in producing valuable, usable products if it has the opportunity to learn together from each customer interaction.

Challenging What You Know About Your Market

You know what they say: to break all the rules, it helps to first master them. (Variations on this have been attributed to many of history's greats, from Goethe to Picasso.) Understanding how significant a shift in your market might be and being able to recognize the trends that portend it will often require a subtle appreciation for the dynamics of the market you compete in. Being able to recognize those shifts early on and distinguish them from meaningless background noise is a characteristic of what they call a "soft mind"—one that has a deep knowledge of a field, but is humble and receptive enough to know that the only constant in tech is change.

Let's give an example: enterprise Software as a Service (SaaS) itself.

Back in the bad old days, we were all told that no "real business" would ever be done with cloud computing. The so-called experts all swore up and down, sometimes (in)famously, that "enterprises" would *never* adopt this software-as-a-service nonsense. We were told that you just *couldn't* do "real enterprise workloads" in the cloud. It sounds bizarre now, but even as late as the early 2010s, we remember old established types (the ones who always remind you of the decades of experience they have in the industry) who swore that SaaS as a model "just wouldn't work" for the enterprise. We won't name names here to save those otherwise very smart people some embarrassment. But those folks had probably lived inside a certain technology and business model for so long that they just couldn't imagine it working in a fundamentally different way, and they weren't very interested in trying. For better or worse, they've been proven wrong now.

It's worth constantly revisiting your premises about the market you compete in, and probably just assuming that a few of them are always incorrect or incomplete—you just don't know which, yet. Many of the business use cases and niches that enterprise SaaS has grown to conquer have actually been areas that earlier or even existing software already served. Part of this is a shift in the capabilities and cost models that SaaS provides, and some of it is probably generational. What we've begun to see more recently is newer, next-wave SaaS solutions attacking entrenched ones with functionality tailored for specific uses or industries. Nimble incumbents are acquiring and expanding feature functionality at a rapid clip to keep up.

In other words, the product managers still have an awful lot to do.

Summary

Industry knowledge is arguably the most important type of domain expertise for an enterprise product manager to develop. Although there is nearly endless content available out there to consume about many industry verticals, only with hard-won experience does the product manager begin to understand what is signal and what is noise. There are many forms of industry research to begin with, but it is critical to inform those with customer feedback. Interviews, shadowing, surveys, and more—the forms of feedback product managers can take from customers (and users!) is nearly endless. This is the kind of *primary* industry research that eventually makes product managers unique experts on the industry they're building products for, and which can't be learned on any third-party research report.

PM Profile

Name: Mike Sukmanowsky

Title: VP of Product, Parse.ly

What have been the most difficult parts of your job to learn?

Going from problem solving to vision setting. In the early days, we focused on finding solutions to problems that people cared about and would pay money for. The good news was, there was no shortage of those problems. The bad news was that not all of those problems were connected to one another. We'd get feature requests that made complete sense from our customers' point of view, but seemed like they were aligned along two different axes of product vision. With a small team, you can ignore this tension for a bit since capacity is limited to one feature at a time and you need to move fast. But as you get larger and unlock the capacity to advance on multiple fronts, this tension boils up to the surface again. We realized we didn't have a great framework for evaluating which of these disparate requests we should accept and started to ask harder questions: what future do we want to create? What's the reason that we exist? What things do we believe that no one else does? I don't know if most adults can answer these questions, so it isn't surprising that companies can also find this a difficult exercise.

What are the most important three skills for an enterprise product manager to have, and why?

Navigating an enterprise requires a bit more skill specialization than a product manager inside of a smaller organization. Above all else, an enterprise product manager needs: communication, scope management, and business acumen.

Communication is a requirement for any product manager, but the number of people inside enterprises makes this skill all the more important. A 15-minute catch-up over coffee with the head of engineering is equally as important as presenting a roadmap to senior management because both can, if uninformed, stall your progress.

Enterprises often come with bureaucracy and legacy systems which can make it difficult to follow Facebook's adage to "move fast and break things." Skilled enterprise product managers can navigate these complexities by boiling a solution down from one massive in scope to the simplest version that still solves the problem and delivers value. Another way to view this scope management skill is that it helps to "sell" an idea inside the organization. If the system rewrite is unpopular because of a perception of low benefits relative to costs, a product manager can engineer a smaller-scale step improvement, backed by data, that helps to build the case. Good product managers are masters in quick wins and ensuring momentum.

What are you looking for when you interview a prospective product manager?

Beyond the usual "design, tech, and business" skills intersection: personal interest in the problem space and listening and communication.

Experience building a similar product is certainly not needed, but interest in the problems our product solves and the customers we serve is. Did the candidate do research on the industry and our customer base? Do they have opinions about where things are going? Better yet, do they have great questions about why we're doing the things we're doing? This initial spark of interest is the base requirement because, frankly, it shows you care. When you already work with a team of passionate people who want to solve tough problems, the last thing you want is to work with someone who just views this as a job to level up their experience. If you're about to interview for a PM role and don't find yourself genuinely interested in what the company does, take 15 minutes to research their product, customers, and competitors. If you're still bored, move on. There are lots of great companies and products out there.

The Product Managers

Making your way into product management in enterprise software might look different than career paths in consumer software. Similarly, how you think about building and managing a team of product managers takes an enterprise slant in a number of noteworthy ways.

We've written extensively here about how product managers can work most effectively with other business functions like development, design, marketing, operations, sales, and leadership. So now, maybe it's time to turn our attention ever so briefly inward for a bit of navel-gazing about the product management team itself.

Among our goals in writing this book, we wanted to not only help guide aspiring product managers prepare for the role, but also advise experienced product managers on how to grow, scale, and improve their teams. This chapter aspires to do just that.

Getting into Product Management

So, you think product management sounds like a cool gig and you'd like to get into it. Good call! Here's our best advice on what you should do.

Product management should almost never be your first or only job in technology. We strongly advise against aiming for it as a first step into tech. Product management, as a generalist role, benefits enormously from the depth of experience its practitioners can bring to bear from other disciplines. Whether that experience is in crafting compelling customer narratives, empathizing with business challenges, building software, or something else, possessing transferable skills positions a product manager candidate to provide substantially more value on day one than a candidate without them. Effective salespeople, savvy marketers, brilliant engineers, and many more can all become first-class product managers, and in our experience, it's very difficult to say that any one group is any better than another.

If you're new to tech, and you're thinking that product management is your goal, a good strategy to get there would be to spend some time in another, adjacent function learning the ropes first. If you're technically inclined, developer or operations roles, or even sales engineering or consulting gigs are great opportunities; if you're not, great marketing and sales talent is always in demand, too. As we've discussed earlier in the book, you don't need to be "technical" to be successful as a product manager, though a certain level of facility with the technology is usually warranted. Spend some time learning, and failing, and then learning some more (don't worry: we've all been there!) in, for instance, product marketing or in pre-sales. The more you learn about your market, the companies within it, customer and user problems, and, of course, the product in question, the more valuable you'll become, both to your current company and to others in the same segment.

Beyond that, the standard-issue career advice applies: be clear with yourself and your manager about your career goals, openly discuss them, and when you feel like your learning curve is beginning to level off—at least a few years into the role—consider making a switch.

How Do Enterprise Product Managers Spend Their Day?

This is one of those questions that product managers of all stripes sort of dread, because there is no really good answer. The nature of product management usually means that every day is different. Of course, there are standing meetings and other recurring things, but instead of structuring a "day in the life," we thought it might be more illuminative to suggest a list of things that most enterprise product managers should aim to accomplish *every two weeks*.

- Stay on top of development status

    — Review what your development team is working on every week. Ensure that all stories in the queue have the right acceptance criteria and that your next upcoming sprint is planned. Connect with the folks on this team at least once every few days at a nondisruptive time.

    — At least once every two weeks, have a standing meeting with your development manager. This can be a quick one-on-one or more in-depth as necessary, but make sure you're talking through outstanding issues, the current sprint, and planning considerations for the next one.

- Get input from Sales

  — Always keep channels with your sales team open—if you're not getting instant/Slack messages, emails, or calls from sellers on a regular basis, you might not be available enough (or, worse, they don't find you useful yet). Pick on a seller every few days and just informally see what's up. Ask what they're seeing and to whom they're talking.

  — Join meetings with the sales teams relevant to your product. You don't need to be there for the entire thing, and joining remotely is fine, but showing the flag and making yourself available is important.

- Gather customer feedback

  — As we discussed in Chapter 6, this is sometimes more easily said than done. You certainly don't need to make a customer visit every two weeks (though if you do, that's great!). But you *can* check on questions coming into customer support and lend a hand if you have time. If customer success is having any calls with customers, they will often welcome product management participation, too. Is there an implementation underway that you can observe? Is the services team working on a project with a customer you can drop in on?

  — You might not be able to get product usage data every two weeks (though again, if you do, kudos!). But if you can, definitely set aside time to review this.

  — If it's relevant for you (and it probably is), check in on industry-relevant social media. Are customers, or potential ones, mentioning your product? A competitor?

- Synchronize with marketing

  — We make a big fuss in Chapter 4 about the importance of working with marketing. Well, we mean it. As a product manager, you should be most interested in what types of prospects marketing is connecting with, and via which channels; who is showing up to your events, and what business roles do they hold; what messaging do they see is resonating, and what is not?

  — Make sure you are aligned regarding content and collateral generation. This can turn into a real mess unless both groups are clear on roles and responsibilities. Ensure that marketing isn't waiting on product management for anything outstanding you don't know about, and review the upcoming content calendar.

  — Perhaps less often—more like once each quarter—stand back from tactical-level detail and ask your marketing colleagues about how *they* see your customer, based on their own view of the market. Does it differ from PM's? If so, interrogate this closely.

- Reserve time for industry knowledge. This might mean catching up on broad trends in your market, analyst research, competitor marketing, and so forth. There is probably far more out there than you have time to consume. Don't try to drink the whole firehose—be selective.

- Work with the Product team. Chances are good that you work with a team of product managers, whether on one large software solution or a platform with many modules. Whatever it is, your entire immediate team should meet together at least once each week. You will doubtless have much to go over in terms of planning, coordination, and sharing information, but always keep these meetings focused on a prearranged agenda. End on time. Everyone is busy.

- Think and plan. Look, our work often acts like a gas: it expands to fill any available volume. If you don't purposefully block out time to think, either quietly and alone or in brainstorming with colleagues (whatever works for you), this critical cognitive work is often the first thing to fall off the plate. Don't let it. Schedule time for deep thinking about your product, your customers, and the problems you are solving. Think about what medium- and long-term problems you want to solve and how you could solve them. This is where the synthesis of your industry, organizational, and product knowledge must come together. Don't rush it, but do not skip it.

How to Know Whether You're Still Growing

One good rule of thumb—and this applies to people in almost any role, not just product managers—is to pay attention to whether you're still growing within your role. If you're not, it might be time to make a move. Here's what we mean.

As any athlete can tell you, growth comes only through struggle and exerting yourself. When you're getting stronger, faster, and better at something, you find yourself constantly challenged. You're constantly having to learn and absorb new bodies of knowledge, apply them in new and different ways, and exercise skills that might start out weak. Sometimes, this will mean that you fail or need to ask for help. (It's okay to do that, by the way.) As you mature in the role, it becomes easier, and you're able to take on bigger, more complex challenges.

But we've all had jobs where the learning curve just sort of levels out. When that happens, you're no longer feeling that challenged. You find yourself uninspired and begin to become bored. This is a tough place to be in. If you're mid-career, and professional progress is important to you, you need to find a new path forward. This might mean moving into product management itself from whatever you did before, or if you're already there, it might mean taking on a new product or set of responsibilities. In some cases, it might mean moving to a new company. (By the way, research shows over and over that switching jobs is also, by far, the best way to increase your compensation.)

On the other hand, keep in mind that frustration and occasionally even resulting boredom are also natural byproducts of any long, difficult, and complicated project. You certainly don't want to use them as an excuse for quitting something mid-stream. All of us have gone through the multistep work project that just wouldn't die. Trust us, see it through. It will continue teaching you stuff long after the project itself is in the books, and you won't burn any bridges by tapping out early.

For Managers: Building Your Team

If you are in the position of leading your own product management team, there are a couple of questions that are (or should) constantly be top of mind. Here are some of the biggest ones we hear.

Spotting and nurturing new talent is a priority for any good manager. In product management, knowing what to look for in new candidates can be essential. Although many different backgrounds can be successful in this role (you know your company context best), we've consistently found a couple of attributes to be shared among successful new product managers:

- They come from a few years of experience in technology already; preferably in the same industry; and even more preferably with the same product or product group. As we said in the previous section, internal functions like engineering, marketing, and sales are some of the richest pools of talent development you have to pick from. (Both of your authors came into product management from functions like these.)

- They are demonstrably passionate about the product and have shown a willingness to go above and beyond their formal job responsibilities to learn more about it. Think about the marketer who becomes deeply adept at using the product, or the engineer who becomes involved in sales cycles, or the seller who acts as their own sales engineer. Each of these people will bring with them a cross-functional set of skills and perspectives into building the future of that product.

- They understand the broader industry landscape, both that which your company sells into and that for software vendors in your space. This isn't just knowing who your competitors are (though knowing the competition well is very handy)—who are adjacent vendors to your solution? What is your Total Addressable Market (TAM)? Have they thought deeply about the broader market challenge for the product?

- They have excellent communication skills. A good product manager needs to be able to communicate very effectively, in multiple formats (written, spoken, nonverbal) to different audiences. They need to be able to talk with ease with engineers, salespeople, executives, and customers, among others.

- They—excuse the expression—give a shit. There is just nothing that can really replace internal motivation for the job, whether you want to call it passion, fire in the belly, or something else. In product management, as in life, there is just no substitute for giving a shit. This person needs to show up ready to play ball. If they do, that's a person you want on your team.

Scaling up your product management team involves hiring the right candidates, of course, but it also involves complex organizational decisions about who covers what, who works with/reports to whom, and how product managers are aligned. Many organizations stick with a straightforward alignment of one product manager to one product, and that often works. Sometimes, as with enterprise Software as a Service (SaaS) platforms that might include numerous discrete products, it might make sense to align more than one product to a single PM.

In other cases, in which one extremely complex product is far too much for just one product manager, dividing up responsibilities between multiple cooperating product managers who report up to one group manager involves clear delineations of responsibilities. (Just ask Ben!) This gets deep into context-specific judgments that are difficult for a book like this one to give broad guidance about; but generally, some of the issues to be carefully considered here are how product knowledge is transferred from one person to the next; who works with development teams relevant to that product, how, and with what bandwidth; and where the product in question is on its maturity curve.

Managing Product Managers as People

Of course, product managers are people, with their own unique emotions, ambitions, insecurities, and more. Anyone, in any role, can be difficult to manage. However, in our experience, good product managers are among the most intrinsically motivated people in enterprise technology, and therefore in some ways the easiest group of talent to manage, all other things being equal. With all but the very least experienced product managers, micromanagement will typically not be necessary. A more seasoned product manager will have had many experiences working with diverse teams in the enterprise, and will mostly be in need of someone to clear a path and act as a force multiplier. Even an experienced product manager (but especially a newer one) will need your help to find the right individuals elsewhere in your company to help grease the skids for a project or feature he is working on, and, as a manager, you should seek to cultivate relationships with managers of other teams so that you can make those connections for your product managers when needed.

Good product managers tend to be happiest when working on projects that bring out their passion. In an enterprise software organization, it can be difficult or impossible to always align your product managers with the projects that they care about. There is simply too much going on across too many teams for everyone to be doing what they love all the time. (As Blair likes to say, the best job you'll ever have, you'll only love about 70% of the time.) Work with your team members to understand what areas of the product or business light their respective fires, and do not be afraid to pair up multiple product managers on a single project if it gets a product manager involved with something she loves. An engaged product manager multiplies the efforts of those around her. Also, with the typical enterprise product manager being pulled in a thousand directions at all times, having double or triple coverage can help. Because product management is by its very definition a collaborative role, the best product managers will appreciate having colleagues who can cover for them if they need to step away.

Product management caters to highly analytical people who want to see others succeed. Of course, not every good product manager is that way, but many are. Because we, as product managers, are in a state of nearly constant assessment of customer problems, product design, and more, it is easy for us to shine the spotlight on ourselves and wonder introspectively whether we are, ourselves, a feature worth keeping around (to borrow a phrase from elsewhere in this book). One day we might feel like we are great product managers and know exactly how to be effective leaders, and the next we wonder how we are even still employed. We often refer to "the sine wave of product management," evoking the frequent oscillation between feelings of success and accomplishment and feelings of discouragement and bewilderment that many good product managers experience. As with any employee, be sensitive to how the team around you is feeling at any given moment, and be aware that product management can be an especially fickle practice for even more senior individuals.

For the Product Manager: What's Next?

If you skipped to this section, we're probably going to tell you what you already know: product managers are generally in high demand in a variety of adjacent functions, most notably executive leadership. As many software companies evolve toward a "product-first" orientation (particularly in SaaS, for reasons we might elaborate on in a future blog post), product management is increasingly becoming one of the most high-visibility roles in the field. "Product" selects for, and then doubles down on, intensively cross-functional operational experience that develops many skills that are difficult to obtain in other roles.

That said, many people either aren't ready for or just aren't interested in climbing the ladder into executive leadership. That's why many product managers we know who transitioned on chose to pursue more specialized paths. Some product managers capitalize on their technical curiosity and go into development roles. Ex-product managers tend to make terrific sales engineers, where they can put to use the customer-facing skills they've developed alongside a strategic, business perspective paired with down-in-the-weeds technical interest. Marketing leadership is another popular path which builds on product managers' strengths in communication to translate product business value to customers. (As we've said earlier, the skills to do this well are inimitable and really quite rare.)

Leadership *within product management itself* is yet another option. One thing to keep in mind about product management leadership roles is how different they are from the front-line product manager gig. Product management leadership is much more about managing humans and solution-level capabilities and value. It involves significantly more prioritization at the 10,000-foot level across different product *areas* (instead of products themselves) and frequently entails much more saying "no." This can be difficult for some people, but it also brings with it significant rewards. Senior product leaders with demonstrated track records of managing successful enterprise products are in short supply and high demand, a situation that seems unlikely to change very soon.

Last Thoughts

We're aware that there is no shortage of generalized, high-level product management advice out there, so we've tried to keep ours as functional as possible up until now. To get all philosophical for just a second, though, just remember that the product manager role is often set up, deliberately, to fail fast and often. Ours is an operational role in which, if you're doing it right, practitioners will routinely run into technical, financial, strategic, or other issues that they don't yet completely understand. This can be very frustrating, and is compounded by the fact that product management relies on the collaboration with other teams to ultimately be successful.

If you feel in over your head, take a deep breath and relax. That means you're probably in the right place to grow. Make sure you maintain a "soft mind," take issues one at a time, and focus. Drink lots of water. Stretch. You're going to do fine.

Summary

If you're looking into getting into product management, make sure you have adjacent skills from within the tech industry—or even better, in the same industry as the one you want to build products for, or at the same firm. Product management should not be a first gig in the software industry—this is a highly cross-functional role, which benefits greatly from using prior domain experience. Product management leaders should always be on the lookout for promising new talent. The best candidates show passion, attention to detail, domain expertise, and an ability to consistently self-motivate. If you're in a product role today, think about where you want it to take you, and be sure you're designing your job to meet those goals.

Conclusion

Well, that's it. That's what we know about enterprise product management. Kinda.

As we said at the very beginning, we set out to describe what we believe are the ideal principles that could be applied to most enterprise software businesses. Above all else, *please don't become discouraged* if your company seems very far from what we've described here! Product management is difficult enough without a couple of book authors telling you that you're doing it all wrong. These are ideals, and they're far from reality even for many of the best product teams, so whatever you do, don't add the burden of perfectionism to your role. You can still do great things to delight and empower your customers. You can still become a great product manager for enterprise software. And you can help your company in big ways by slowly, carefully implementing some of what we've described here. Change often happens slowly in enterprise software, and our goal here was to give you a few places to begin.

One of the most difficult parts about writing this book was trying to distill all the sprawling branches of our respective experiences into well-organized, consumable sections that we hoped you would find easily usable. In doing that streamlining, we left out a lot of stuff that just didn't fit anywhere very cleanly. We also ran into even more enterprise product managers whose stories and experiences we wanted to share in profiles like the ones you've read here. So following publication, we plan to post more content as one-off blog posts at the book's website, BuildingForBusiness.com (*http://www.buildingforbusiness.com/*). We hope you'll check that out to see what else we came up with.

Something that surprised both of us while we wrote this book was the contrast between some of our experiences. Mature versus immature products; market leaders versus scrappy underdogs; big companies versus small ones; different organizational and company practices. The practice of product management truly does vary widely across many parts of the enterprise software industry, which makes it a really exciting place to work right now. The rules are still being written; the boundaries are still extremely fuzzy. Accordingly, we're keenly aware that a lot of people might not agree with all of the views on product management that we've expressed here, and that's fine. Like we said in the introduction, we're still learning about what works and what doesn't. Be very careful when you feel like you've got things all figured out; chances are, you're closer to being left behind.

We wanted this book to make a meaningful contribution to the emerging literature on product management from the enterprise software perspective, which we feel is too often ignored. We hope it's been helpful to you in some way and that you'll let us know how we did.

Now don't you have a backlog to go prioritize?

See y'all out there,

Blair & Ben

@BlairReeves (*https://twitter.com/blairreeves*)

@BenjaminGaines (*https://twitter.com/benjamingaines*)

Index

About the Authors

Blair Reeves is a Principal Product Manager at SAS Software. He has previously held product and marketing roles at Demandware (now the Salesforce Commerce Cloud) and IBM. Blair is an avid hiker, writes and codes creatively, and believes inspired products are built first in the gym. He writes about enterprise software, product management, and wide-ranging topics at his blog, *Blair-Reeves.me*, and has an annoying Twitter habit at @BlairReeves. He and his wife live with their daughter and dogs in Pittsboro, North Carolina.

Ben Gaines is a Group Product Manager for Adobe Analytics Cloud. In this role, he works closely with Adobe customers to understand their needs and guide the product strategy to address their most pressing problems. A native Bostonian transplanted to Salt Lake City, Ben plays basketball compulsively, and also enjoys cooking, tourism, and playing with his four kids. You can consume Ben's every waking thought on Twitter at @benjamingaines.

Colophon

The animal on the cover of *Building Products for the Enterprise* is the *tule elk*.

The tule elk is found only in the U.S. state of California, ranging from the Central Valley out to the coast. There were believed to be around 500,000 tule elk when the Europeans arrived in California. Almost driven to extinction during the Gold Rush of the early 19th century, the wild population now numbers around 4,000.

It is the smallest of the American elk, with males weighing 450–550 pounds (200–250 kilograms) and females 375–425 pounds (170–193 kilograms). They are about midway in size between the larger reindeer and the common white-tailed deer. Full-grown bulls (over 2 years old) have antlers with 5–7 points that shed in early spring (that is, after rutting season) and regrow over the summer.

The tule plant on which this elk feeds is a tall, grassy brush that grows in the marshes and lakefronts of the California coast. The indigenous peoples of the area used tule to make sleeping mats, canoes, and even shelters.

Many of the animals on O'Reilly covers are endangered; all of them are important to the world. To learn more about how you can help, go to *animals.oreilly.com*.

The cover illustration is by Karen Montgomery, based on a black-and-white engraving from *Meyers Kleines Lexicon*. The cover fonts are URW Typewriter and Guardian Sans. The text font is Scala Pro, and the heading font is Benton Sans.

Learn from experts.
Find the answers you need.

DISCARD

CPSIA information can be obtained
at www.ICGtesting.com
Printed in the USA
BVOW11s2043040318
509596BV00014B/165/P